"BENT WINGS

F4U CORSAIR ACTION & ACCIDENTS

True Tales of Trial & Terror!"

"BENT WINGS

F4U CORSAIR ACTION & ACCIDENTS

True Tales of Trial & Terror!"

by

Fred "Crash" Blechman

. . . and others

3-BLEC

(excerpt from Chapter 1)

I checked my altitude by seeing where the clear, flat horizon crossed the ship's mast above the bridge, since that indicated exactly how high I was above the deck. At approximately the 90 degree position on the base leg I picked up the LSO with his colored paddles on the port fantail. Now the challenge was to keep the ship from getting ahead of me, since it was churning away from me at roughly 60 feet per second (including the surface wind that was trying to drag me even further behind). I watched the horizon crossing the bridge for altitude, and carefully controlled the power and nose attitude for holding around 90 knots—just a few knots above stalling!

I used a simple technique to properly intercept the ship. I put the left side of the Corsair's nose on the center of the deck at the aft end—and held it there! If I tried to judge my turn any other way I would invariably get sucked back behind the ship with a straightaway to catch up—but then I'd lose sight of the LSO under the Corsair's long nose.

There was no luxury of any significant straightaway in landing on those old straight-deck carriers when you were flying a long-nose Corsair in a nose-up attitude. You just couldn't see ahead of you—only off to the side. We essentially pyloned counter-clockwise around the LSO in order to keep him in sight!

As I got close in, I tried to keep the nose aimed toward the ship's centerline. This was not only affected by the ship's forward motion, but also by the wind over the deck. This wind was seldom straight down the deck, but approximately 15-degrees to port so the turbulence from the ship's stacks and bridge did not appear in the flight path of the landing planes. This made for a very tricky approach and last few seconds. . .

At this slow speed, just a few knots above stalling, it took a lot of right rudder, even though in a left turn. And you didn't dare add power quickly since the powerful engine turning that large prop could make the aircraft roll uncontrollably to the left—the dreaded "torque roll."

It took a lot of back stick, considerable power, and right rudder to hang in there. As I approached the ramp in a left turn, the LSO's paddles and my own perception was that I was drifting to the right of the deck centerline. Too much right rudder. I cross-controlled a bit and slipped to the left just as I approached the ramp, and got a "cut," the mandatory command to cut my power and land.

"Ah, landing number 5," I thought as I relaxed, dropped the nose, and pulled back to drop the tail so my hook would catch an early wire. But I relaxed too soon!

3-BLEC

Library of Congress Number: 98-83109
ISBN#: Hardcover 0-7388-0345-6
Softcover 0-7388-0346-4

This book was printed in the United States of America.

To order additional copies of this book, contact:
Xlibris Corporation
PO Box 2199
Princeton, NJ 08543-2199
USA
1-888-7-XLIBRIS
1-609-278-0075
www.Xlibris.com
Orders@Xlibris.com

CONTENTS

PART VI

Wallace Bruce Thomson

PART VII

by John R. "Jack" Eckstein, CAPT USN (Ret.)

PART VIII

Miscellaneous Tales of Trial & Terror!

DEDICATION

THIS BOOK IS DEDICATED TO ALL FORMER
F4U/FG CORSAIR PILOTS, NAVY AND MARINE,
U.S. AND FOREIGN, DEAD AND ALIVE,
WHO HAD THE THRILL OF FLYING
ONE OF THE STRANGEST-LOOKING, POWERFUL AND
SUCCESSFUL PROP-DRIVEN FIGHTER PLANES
OF WORLD WAR II AND KOREA
—AND TO THOSE WHO WISH THEY HAD!

PREFACE

by CAPT. 'Ham" Tallent, USN

CAPT Hamlin "Ham" Tallent, Director, U.S.Navy Aviation Officer Distribution, in "Wings of Gold," the official magazine of the Association of Naval Aviation (Fall, 1998) wrote an article about why Naval Aviators and Flight Officers joined the Navy in the first place.

The following, which seems so appropriate to this book, and the days past as well as now, are excerpts from that article (with permission):

You and I joined the Navy because we wanted to be warriors. I know that sounds corny, and maybe some of you wince at the word. I'm not ashamed to say it for us. We joined the Navy because we wanted to be warriors. I don't mean we wanted to crush skulls and drink blood. We wanted to be a member of something special. We didn't want to be just some guy on the street. We wanted to be better than we were and we wanted to join an organization that would help us be better.

We wanted to be part of something noble and good and glorious. We wanted to be part of something that made our family smile with pride when they mentioned our name and explained why we weren't home for Christmas. We wanted to do something that made our dad go out of his way to introduce us to his old friends. We wanted to be warriors.

We wanted to do something not everybody could do or even wanted to do. We wanted to be part of something we could be proud of forever. Not just the heady, leather-jacket years of our youth but more importantly, during the frustration of our majority. We wanted to be something that people of honor and courage regarded as precious. We wanted to be part of something grand that linked us with the history of this country and something fearsome that would preserve its future. We wanted to be warriors.

By far, the thing we wanted most was to be part of something that allowed us to live as much with our hearts as with our heads. We didn't want to be part of something totally rational or ruled by a bottom line. We didn't care if the ledger added up or if the columns were in a row. We didn't want our lives ruined by too much pragmatism. We picked pride and passion over wealth and even well-being and we rejected the notion the individual was more important than the whole. We might have claimed we joined because we wanted to fly or because we wanted to travel, or to have adventures or maybe because we didn't have anything else to do. We might have said these things, but the truth is, we wanted to be warriors.

FOREWORD

You are in an F4U Corsair, turning toward the carrier and preparing all the controls for a night carrier landing. You are dive-bombing the Japanese battleship Yamato. You are flying in the clouds on instruments—and they seem to be lying to you. You are in combat and watch your best friend get shot down. . .

These are just a few of the stories in this book, written by the Corsair pilots who lived to tell the tales. All of these stories are true to the writers' best recollections and reference material, such as pilot logbooks, squadron history, and accident reports.

If you've ever flown a "bent wing" Corsair, you'll find stories here that will bring back memories, and you probably have your own "true tales of trial and terror!"

Or perhaps you're one of the many "wannabees" that were born too late to fly the World War II warbirds, and you'd like to experience some vicarious thrills by reading about those who did.

The stories in this book—all written by me or former Corsair pilots I personally know—are bits and pieces of recollections from World War II and the Korean War period, when the Chance-Vought F4U and Goodyear FG Corsairs were considered the premier propeller-driven fighters and attack aircraft. Some Corsair history can be found in APPENDIX B.

Each of these stories is relatively stand-alone, not depending on previous stories, so some facts may be repeated. No particular chronological sequence is used, so you can pick up the book, flip it open and read any "tale" without wondering if you've missed an important earlier fact.

And if you like to read accident reports, there are four of my own (following Chapters 1 and 13), and 19 others in APPENDIX A, all from my Navy fighter squadron (VF-14 Tophatters) in the two years I was with them—which gives you an idea of how treacherous the Corsair could be!

Then, if you are into electronics or computers and want a break from "Trial and Terror!" you might find some humor in APPENDIX C—which has nothing whatever to do with Corsairs or flying!

Fred "Crash" Blechman
West Hills, California, November 1998

PART I

Fred "Crash" Blechman

Fred Blechman, originally from Far Rockaway, New York, joined the Navy V-5 program in July, 1945. As an Apprentice Seaman he attended Bethany College in West Virginia, Swarthmore College in Pennsylvania, and Columbia University in New York before reporting to Dallas, Texas, as an Aviation Cadet (AvCad) for Selective Flight Training in the N2S Stearman.

He soloed in an N2S on 16 September, 1946, and was sent to Ottumwa, Iowa for Pre-Flight. However, the Holloway Plan intervened, and Fred left the Navy to study Aeronautical Engineering at Cal-Aero Technical Institute in Glendale, California.

He re-entered flight training as a Naval Aviation Cadet (NavCad) in November, 1948. After Pre-Flight, Basic Flight Training, and Carrier Qualification (CQ) in SNJs at Pensacola, he went to Corpus Christi, Texas, for Advanced Flight Training in F4U-4 Corsairs. He earned his Naval Aviator "Wings of Gold" on August 23, 1950.

Fred flew F4U-5 Corsairs with the VF-14 "Tophatters," home based in Jacksonville, Florida, and made two Mediterranean cruises with the squadron before leaving active duty in November, 1952. After a short time in the Navy Reserve, he left as a LT.

He completed his Aeronautical Engineering studies and worked in the aerospace industry for 15 years until he started his own business. Since 1961 he has written more than 750 magazine articles and seven books about electronics, microcomputers and flying.

He is now retired with his wife, Ev, in West Hills, California.

CHAPTER 1

"F4U Corsair Carrier Qualification"

by Fred "Crash" Blechman

Finally, after 13 years of dreaming about becoming a Naval Aviator and earning my "Wings of Gold," this was my "final exam." Making six arrested carrier landings in an F4U-4 Corsair would earn me my gold wings and Ensign's commission. I had no idea I was about to crash.

It had been almost 21 arduous months since I had entered flight training. I had over 200 hours in SNJs, six arrested carrier landings in an SNJ, then almost 100 hours in Corsairs. Now, getting ready for Corsair carrier qualification, I had made 91 field carrier landing practice (FCLP) approaches and landings at Bronson Field near Pensacola. Just six carrier landings in a Corsair and I would "graduate."

So here I was, at about 9AM on August 10, 1950, flying F4U-4 Corsair #80893, together with five other students and our instructor, heading out to our carrier in the Gulf of Mexico off Pensacola. We rendezvoused with the light carrier U.S.S.Wright (CVL-49) as it churned at approximately 25 knots through the waters near Pensacola, Florida. The sea was calm with only occasional whitecaps from the gentle breeze. The azure sky was punctuated with random cotton balls. All was serene. Life was good. This was the day I'd been waiting for through so many episodes of "trial and terror."

Our flight received a "Charlie" landing clearance, formed a

right echelon, and streaked upwind by the starboard side of the ship at about 800 feet as we peeled off to establish our landing intervals.

This was busy-time. Wheels, hook, flaps, power settings, trim, setting the beam position and interval while headed downwind, turning toward the carrier at the proper position, losing altitude, losing airspeed, spotting the landing signal officer (LSO), responding to LSO signals, adjusting bank and nose attitude. . . busy, busy time.

This was the real thing. There was no way we could accurately simulate landing on a moving carrier with those FCLP hops at Bronson Field—but they were the best means available to practice flying low and slow, follow the LSO's signals, and set the proper speed and attitude for a carrier approach in the "Hose Nose" Corsair.

My first four landings were normal, with no waveoffs, as we each in turn made our landings and takeoffs. After catching a wire, the barriers were dropped, and we made a deck-launched takeoff. But I was getting tired, and my light summer flight suit was drenched with sweat. I had no way of knowing that the next landing, #5, was going to be very different. . .

"Only two more landings to go," I thought as I prepared for my deck launch. With a ten-knot surface wind and the carrier's forward speed, the wind over the deck was approximately 35 knots. The takeoff should be easy. I checked various settings. Full flaps. Cowl flaps open. Hook up. Trim 6 degrees nose right, 1 degree nose up, 6 degrees right wing down. Tailwheel locked. Cockpit canopy open and locked. Shoulder straps and seat belt tight. Prop control full forward for maximum revolutions per minute (rpm). Mixture auto rich. Supercharger neutral. Wings locked. Controls move freely.

I watched the Launch Control Officer to my right give me the windup signal with his right arm as he pointed to my engine with his left arm. I advanced the throttle to 42 inches of manifold pressure and applied full toe brakes by pressing down the tops of the rudder pedals. At above 44 inches the wheels would start slipping on the deck, so full power could not yet be used. I held the joystick all the way back to keep the tail from lifting up and possibly digging the tips of the 13-foot four-bladed propeller into the wooden flight deck.

The 2100 horsepower Pratt and Whitney R-2800-18W(C) Double-Wasp 18-cylinder radial engine roared and the whole airplane shook with anticipation as I verified proper engine readings and signaled I was ready with a head nod. (I dared not let go of the stick for a right hand salute, or the tail could come up!) The Launch Control Officer threw his arm forward with two fingers extended, the signal for me to release the brakes and take off.

Surging forward, the Corsair picked up speed and rumbled down the deck. I added throttle to full power—approximately 54 inches of manifold pressure—and held a lot of right rudder to counter the torque of the huge engine and propeller sticking out 15 feet ahead of me. Releasing back stick pressure, the tail lifted and I could finally see where I was headed. I aimed for the right side of the deck, lifting off easily before the ship slipped behind, with nothing but rippling water beneath me. A slight right turn cleared my slipstream from the plane landing behind me, as I climbed ahead of the ship at 125 knots to the 800-foot pattern altitude. Since I was just going around to make another landing, I left the flaps and wheels down. At pattern altitude I reduced the throttle setting to 34 inches of manifold pressure, set the propeller to 2300 rpm, and reset the trim tabs for neutral stick pressure.

About a mile ahead of the ship I made a 180-degree left turn, descending to 200 feet for the downwind leg. I dropped my tailhook, unlocked my tailwheel, and set myself up approximately 3000 feet abeam of the ship, fast approaching on my port side as it steamed upwind.

Landing #5

The plane was flying smoothly with the canopy open and locked. The hot Gulf air and the roar of the engine blustered in from both sides of the windshield. Everything in the cockpit seemed A-okay, warm and comfortable as an old shoe as I watched the ship slip past my nose and toward my left wing.

As the straight deck of the light carrier Wright steamed upwind and its wake appeared ahead of my left wingtip, I banked sharply toward the ship's stern and began slowing the airplane down to an approach speed of 90 knots. Check flaps down, wheels down, hook down, tail wheel unlocked. I shoved the prop control forward for full rpm and reset the trim tabs to takeoff settings in case of a waveoff. I set my rate of descent to about 150 feet per minute, maintaining just enough throttle to hold the nose up approximately 15 degrees, hanging on the prop.

I checked my altitude by seeing where the clear, flat horizon crossed the ship's mast above the bridge, since that indicated exactly how high I was above the deck. At approximately the 90 degree position on the base leg I picked up the LSO with his colored paddles on the port fantail. Now the challenge was to keep the ship from getting ahead of me, since it was churning away from me at roughly 60 feet per second (including the surface wind that was trying to drag me even further behind). I watched the horizon crossing the bridge for altitude, and carefully controlled the power and nose attitude for holding around 90 knots—just a few knots above stalling!

I used a simple technique to properly intercept the ship. I put the left side of the Corsair's nose on the center of the deck at the aft end—and held it there! If I tried to judge my turn any other way I would invariably get sucked back behind the ship with a straightaway to catch up—but then I'd lose sight of the LSO under the Corsair's long nose.

There was no luxury of any significant straightaway in landing on those old straight-deck carriers when you were flying a long-nose Corsair in a nose-up attitude. You just couldn't see ahead of you—only off to the side. We essentially pyloned counter-clockwise around the LSO in order to keep him in sight!

As I got close in, I tried to keep the nose aimed toward the ship's centerline. This was not only affected by the ship's forward motion, but also by the wind over the deck. This wind was seldom straight down the deck, but approximately 15-degrees to port so the turbulence from the ship's stacks and bridge did not appear in the flight path of the landing planes. This made for a very tricky approach and last few seconds. . .

At this slow speed, just a few knots above stalling, it took a lot of right rudder, even though in a left turn. And you didn't dare add power quickly since the powerful engine turning that large prop could make the aircraft roll uncontrollably to the left—the dreaded "torque roll."

It took a lot of back stick, considerable power, and right rudder to hang in there. As I approached the ramp in a left turn, the LSO's paddles and my own perception was that I was drifting to the right of the deck centerline. Too much right rudder. I cross-controlled a bit and slipped to the left just as I approached the ramp, and got a "cut," the mandatory command to cut my power and land.

"Ah, landing number 5," I thought as I relaxed, dropped the nose, and pulled back to drop the tail so my hook would catch an early wire. But I relaxed too soon! Perhaps I was more tired than I realized, but my wings were not level, and I didn't pull back soon enough. The left main gear hit first, blowing the tire, and the plane bounced back in the air. At this point the tailhook caught the #3 wire and slammed the Corsair back on to the deck. On this second impact the left wheel strut broke and the right tire blew out!

I was thrown with more force than usual against my shoulder harness as the plane tilted to the left and settled on the deck. The carrier crash horn blew. Deck hands, some carrying fire extinguishers, came scampering up from the catwalks and surrounded the airplane. Controlled pandemonium reigned as I was quickly unbuckled and helped out of the cockpit, since fire after a crash was always a danger.

A Corsair zoomed overhead taking a "fouled deck" waveoff. It was Midshipman John A."Jack" Eckstein, my roommate and good friend through most of flight training. He told me later he was so shaken by my accident right in front of him as he was making his approach for his fifth landing that it took him several more passes to get in his last two landings. (He got his wings, stayed in the Navy, and retired as a Captain.)

I was not injured at all—except for my pride. But I was very concerned about being washed out of flight training, shattering a 13 year dream—and with only one landing to go! I had special reason to be concerned since I had my only previous accident just three weeks before when I torque-rolled a Corsair on a waveoff during my first field carrier landing practice flight at Bronson Field, and crumpled the left wing. No personal injury there, either, and a Student Pilot Disposition Board allowed me to continue training.

Disposition Board—Again!

Now I had to appear a second time before the Student Pilot Disposition Board to determine if I would get washed-out, or would get the chance to make that one remaining landing (the crash counted as #5) to get my wings. Was it my unblemished record prior to three weeks earlier, was it my sincerity and obvious strong desire to become a Naval Aviator, or was it the fact that North Korea had invaded South Korea a month or so before, and the Navy was calling up the Reserves and anticipated the need for more pilots? Whatever the reason, I was awarded some additional field carrier landing practice and another try for that last carrier landing!

Five days after the crash I climbed aboard the same Corsair, #80893, now with new tires and a new port landing gear strut, and made five field carrier practice landings at Bronson Field, and was considered qualified to make that last arrested landing needed to get my wings. Three days later, on August 18, I walked aboard the U.S.S. Wright in port at 6AM. The carrier steamed out into the Gulf of Mexico for that day's carrier qualifications.

Landing #6

The first flight of Corsairs appeared at 9AM and began their qualification landings. The first to complete his six landings was NavCad Vince "Rick" Ricciardi, whom I'd known since pre-flight. I congratulated him as he climbed down from his Corsair, #97168, and I clambered aboard. I strapped myself in with the help of a plane captain, checked all the power and control settings, and deck launched. One landing to go.

This was it! If I had too much trouble getting aboard, or crashed again, it was certain I would be washed out. The takeoff and downwind leg were normal, but as I made the approach I got more tense than usual as I considered the consequences of failing. This probably made me concentrate more than in previous landings, since I got a "Roger" flag signal from the LSO all the way into the cut, and caught the #3 wire. I did it! I had qualified to be a Naval Aviator!

The ceremony for commissioning as Ensign, and receiving the "Wings of Gold," was held at Pensacola on August 23, 1950. My mother flew in from New York to pin on my wings and bars. I've never done anything more difficult—or of which I'm more proud—than earning those gold wings! And after over thirty arrested carrier landings, I learned to drive a car. . .

SIDEBAR

Flashback—First Try

I was six years old in 1933 when I went up for my first $5 plane ride over New York City. It left me with an indelible impression of all those little houses, little cars, little roads, plowed fields, and tiny, tiny people—and how the whole world twisted and turned as the pilot maneuvered the airplane. I loved it! However, it wasn't until 1937, at age ten, at a Navy airshow with fat, gray-and-yellow Navy biplanes, that I decided I was going to be a Navy pilot!

After eight years of building model airplanes and devouring flying magazines, my chance came in July of 1945 when I joined the Navy V-5 program as an Apprentice Seaman for four semesters of college training in uniform before entering flight training. Finally, in August of 1946 I became an "AvCad," the term used at that time for Aviation Cadets. After eight flights in an N2S Stearman "Yellow Peril" in Dallas, Texas, I successfully soloed on September 16. Then it was on to pre-flight training at Ottumwa, Iowa.

But World War II was over, downsizing was in place, and we were given the option to sign up as Midshipmen for four more years under the Holloway Plan, or go back to civilian life and complete our college education under the G.I.Bill. I got out.

Second Try

However, I maintained contact with John Higson, who had stayed in the program, and heard about the "Ab Initio" (From the Beginning) program my former classmates were beginning at Cabaniss Field in Corpus Christi. They were starting out in SNJs as the primary trainer instead of the Stearman—and I would have been in the first class to do this! This drove me nuts. I haunted the Navy recruiting office trying to get back into Navy flight training. It took two years, but in November of 1948 I got back into flight training and headed to Pensacola for pre-flight. This time we were called "NavCads," a new designation that officially began on June 22, 1948 with a new Navy flight training program.

I completed pre-flight at Pensacola, then basic flight training in SNJs at Pensacola (with six arrested carrier landings on the U.S.S. Cabot (CVL-28) on 23 March, 1950), advanced flight training in F4U-4 Corsairs at Cabannis Field in Corpus Christi, and then back to Pensacola for Corsair carrier qualification. Oh, by the way, being a city-boy, I had never learned to drive a car, but I was flying Corsairs!

ACCIDENT REPORT #1

Date: 18 July 1950, 14:35

Pilot: NavCad Frederick Blechman USNR

Organization: CQTU 4, NAAS, Corry Field, CNABT, CNAT, BUAER
Aircraft: F4U-4 #62132
Purpose: FCLP
Hrs.last 3 months: 75.1; Total hours: 282.6
Location: Bronson Field
Weather: Contact
Injuries: None

SPECIFIC ERRORS:
Pilot failed to level his wings as he applied power for waveoff. Pilot permitted plane to stall while attempting waveoff.

ANALYSIS:
Pilot was on FCLP syllabus. He was making his second approach of the period and overshot the groove. The LSO waved him off and the pilot added power but continued to turn in towards platform. Plane continued to roll to left and struck runway. Initial impact was on left wing followed by left wheel and then right wheel about 60-feet past LSO platform. Upon impact pilot closed throttle and plane remained on landing gear. It rolled to stop heading approxi-

mately 45-degrees from duty runway about 70 yards from LSO platform.

Pilot attempted to take waveoff from turn after overshooting groove. He made no attempt to level wings until after full power had been applied. Sudden increase in engine torque plus fact that he was already in turn stalled out left wing and made it impossible to return plane to level flight after wing started down because of insufficient altitude.

Pilot after his initial mistake used good judgment after wing struck runway by taking off all power and remaining on ground. This action minimized damage to aircraft.

This accident could have been avoided if pilot had leveled his wings as he applied full power for waveoff.

SPECIAL EQUIPMENT:
Shoulder harness effective.

LOCAL RECOMMENDATIONS:
(1) That Blechman be ordered to appear before Student Pilot Disposition Board. (2) That circumstances leading to this accident be explained to Students and Instructors attached to this unit.

COMMANDING OFFICER:
Student remanded to Student Pilot Disposition Board which awarded him extra time and to continue training. It is believed that this accident was due to lack of experience rather than lack of aeronautical skill. Analysis of accident posted on squadron safety bulletin board.

REMARKS:
Damage: Left wing and aileron.

ACCIDENT REPORT #2

Date: 10 August 1950, 13:42

Pilot: NavCad Frederick Blechman USNR

Organization: CQTU-4, NAAS Corry Field, CNABT, CNAT, BUAER
Aircraft: F4U-4 #80893
Purpose: Carrier Qualification
Hrs. last 3 months: 80.3; Total hours: 303.2
Location: USS Wright (CVL-49)
Weather: Contact
Injuries: None

SPECIFIC ERRORS:
Pilot did not land plane after cut as instructed, but instead nosed over excessively and landed wheels first.

ANALYSIS:
Pilot was making his fifth carrier landing aboard the USS Wright. The approach was good and the "cut" given. After cut, pilot dived for the deck, landing on his port main gear. The port tire blew on impact. The aircraft bounced but engaged the #3 wire, which pulled plane back to the deck. On second impact, the port strut broke and the starboard tire blew. Pilot nosed over excessively after the cut.

SPECIAL EQUIPMENT:
Shoulder harness averted possible injury.

LOCAL RECOMMENDATIONS:
That NavCad be ordered to appear before a Student Pilot Disposition Board. All instructors are cautioned to keep rebriefing the students on proper landing technique.

COMMANDING OFFICER:
Student was remanded to Student Pilot Disposition Board, which awarded him one warm-up and a recheck. Analysis of accident and improper procedures used were presented to all students and posted on Squadron Safety Bulletin Board.

REMARKS:
Damage: Landing gear left and right, propeller.

NavCad Blechman in F4U-4 cockpit, July 1950, Pensacola, preparing for field carrier landing practice.

Ricciardi is out, I'm strapped in, with one landing to go to get my wings. . .

CHAPTER 2

"Earning the Mudhole"

by Fred "Crash" Blechman

A Dream is Born

I was nearly ten years old on Sunday, July 4, 1937 when my parents took me to an airshow at Floyd Bennett Field in New York City—a Naval Air Station at that time. My face was pressed right up against a chainlink fence when a small group of fat Navy silver and yellow fighter biplanes (now I know they were Grumman F3Fs) flew over the field in a right echelon, peeled off, landed, taxied up, and parked no more than 50 feet from me!

I watched wide-eyed as the pilots, with their cloth helmets and goggles and flowing white scarfs, climbed out of the tiny cockpits and clambered down the sides of their chunky fighter planes. I saw them gather together, tall and handsome all, and was thrilled when they ambled over to the crowd at the fence. One of them even talked to me! "Wow," I thought, "I wanna be one of those guys. When I grow up I'm gonna be a Navy fighter pilot!" At that time it was just a dream. . .

I read flying books, built solid balsa-wood models and stick-and-paper flying models, and devoured everything I could find about flying. Throughout World War II I followed the exploits of the flyers, always planning that one day, when I was old enough, I'd join up to fly.

First Try

My chance came in 1945 when I graduated from High School in January and applied for the Navy V-5 program. If I passed the physical and written tests, the Navy would send me to two years of contract college training before any pilot training. I was 17 years old, and not exactly a tall, handsome, muscular poster-pilot type. I was only 5-feet 10-inches tall, weighed only 135# with a slim 28-inch waist, was plagued with teen-age acne—and didn't even know how to drive a car! Nevertheless, desire and determination overcame my shortcomings. I passed the physical, mental and psychological testing and I got orders to report to Bethany College in West Virginia as an Apprentice Seaman for my first V-5 semester in early July, 1945.

We were actually on active duty, always wearing our lowest-of-the-low apprentice seaman uniforms and marching to and from every activity. While I did not find the studies particularly hard, I found the physical activities difficult; constant marching drills and considerable physical training, including swimming, calisthenics, and competitive sports. I wanted to fly, so I endured.

About a month after reporting to Bethany College, on August 6 the atom bomb was dropped on Hiroshima. World War II quickly ended, and the Navy wondered what to do with those of us in the V-5 pilot pipeline—but still in the college training phase. By the time they figured it all out, I had completed the other three semesters at Swarthmore College in Pennsylvania and Columbia University in New York.

Selective Flight Training

It was now Spring 1946, and the Navy was downsizing its need for pilots and preparing to close down the V-5 program. It was decided that instead of the expense of sending all who completed two years of college to 16 weeks of pre-flight training, they would first select those who were at least capable of learning to fly. The weeding-out process took place in late 1946 and was called "Selective Flight Training."

In August 1946 I was made an AvCad (frequently incorrectly termed "NavCad") and sent to NAS Dallas (Hensley Field, Grand Prairie, Texas, halfway between Dallas and Fort Worth) in Class 13-46-C. The requirements were simple enough: solo after eight training flights and a check ride in tandem dual-control N2S-5 Stearman "Yellow Perils," or you wash out!

Finally we were out of our seaman uniforms and into officer-like khakis with distinctive collar anchors and a neat embroidered V-5 cap insignia. We were finally going to fly!

Earning the Mudhole

The eight flights were about an hour each, with instructors not particularly thrilled about their duty. The instructor sat in the front seat. The student in the rear seat wore a special helmet that had rubber tubes extending forward to a mouthpiece in the front cockpit that the instructor spoke (yelled) into; this was known as a "gosport." The student could hear, but not speak! The instructor had a mirror that allowed him to watch the reaction of the student behind him at all times.

It was in this intimidating environment that we went through the eight-hop syllabus: Controls, Climbs, Spins, Take-offs, Landings, Landings, Landings, Finishing Touches. I had no particular problems, as I recall, with the flying, but particularly enjoyed taxiing the plane around, since I didn't drive a car!

After the eight flights, it was time for a check ride by a different instructor. If you got an "up" you were cleared for a half-hour solo flight from the rear seat. Edgar M. "Ed" Housepian, now an M.D. and Professor of Clinical Neurological Surgery at a prestigious hospital in New York City, put it this way in a recent letter: "Soloing was a great experience. I'd never been able to get the plane on the deck through all of my practice hops with my instructor. Finally he said, 'You've tried to kill me enough times. Try it with your check pilot.'

For some reason I greased in three perfect landings with the check pilot and was stunned to find myself going around by myself. About the third time I freaked thinking, 'What am I doing here? I don't know how to fly this plane!' but came down safely."

I started flying on September 3, 1946, and soloed on September 16 at Arlington Field, a Dallas outlying field where we did our practice and solos. My instructor was LT M.K."Mel" Crawford, and my check pilot was ENS E.L."Carp" Carpenter. I recall my first solo flight as one the most thrilling times of my life up to that time. The freedom and exhilaration of being in total control (just push the stick to the side a bit and the whole world tilts!) and the great feeling of accomplishment on completing a worthwhile goal after—for me—considerable adversity. No one was telling me what to do through a one-way gosport, and I wasn't being constantly watched through a mirror. I was on my way to being a fighter pilot!

A common tradition when a pilot completes his first solo is to cut off his tie. But summer rainstorms are common in the Dallas area, and there were lots of muddy holes around the tarmac area. So, the first-solo indoctrination at NAS Hensley Field that summer (in addition to clipping the tie) was to tear off the AvCad's khaki shirt-tail and throw the cadet in a slimy worm-infested mudhole! When I stepped out of my plane at the main base I got my indoctrination. I had "earned the mudhole," and it took two long showers to remove the sticky mud and green worms. Yuk!

But I survived that, and like many of the others, decorated the shirt-tail with colored cartoons, and had the other guys sign it. I still have that shirt-tail. Of the 25 signatures, I have recently located and talked with eight of those "mudholers"—after almost 50 years!

The Holloway Plan

Those of us who soloed were sent, in late 1946, to Ottumwa, Iowa, in the cold, snowy dead of winter for pre-flight and primary flight training—except for a slight change. That's when the Holloway Plan hit us. The war was over, and too many cadets were in the pilot pipeline. We were told we would have to sign up as Midshipmen for FOUR MORE YEARS, with no Ensign commission for two years (even if we earned our pilot wings sooner!)

We were also told if we stayed at Ottumwa through the cold winter, we'd be pushing Stearmans around—tarmac duty—for at least 6 months before getting into actual flight training. Or, as an alternative and inducement to reduce the pool of flight trainees, we were allowed to keep all our neat officer-like uniforms and $200 mustering out pay if we went back to civilian life. Considering my chances were poor of completing flight training with the radical downsizing, I accepted the alternative!

Second Try

However, I maintained contact with John Higson, who stayed, and heard about the "Ab Initio" (From the Beginning) program my former classmates were beginning at Cabaniss Field in Corpus Christi—starting out in SNJs as the primary trainer instead of the Stearman—and I would have been in the first class to do this! This drove me nuts. I haunted the Navy recruiting office trying to get back into Navy flight training. It took two years, but in November of 1948 I got back into flight training and headed to Pensacola for Pre-Flight. This time we were called "NavCads," a designation that officially began on June 22, 1948 with a new Navy flight training program.

I completed Pre-Flight at Pensacola, then basic flight training in SNJs at Pensacola (with six arrested carrier landings on the USS Cabot (CVL-28) on 23 March, 1950), advanced flight training in F4U-4 Corsairs at Cabannis Field in Corpus Christi, and Corsair carrier qualification on the USS Wright (CVL-49) on August 18, 1950.

On August 23, 1950—13 years after I saw the tiny F3Fs at Floyd Bennett Field—I got my Navy "Wings of Gold." I was Naval Aviator #T891. I was a Navy fighter pilot. My dream had come true. . .

I joined the VF-14 "Tophatters" at Jacksonville, Florida (Cecil Field) in September 1950 as junior ensign, flying the latest model F4U-5 Corsair, and made two Med cruises until separation as LTJG in November of 1952. . . and after about 30 carrier landings in Corsairs, another dream came true—I finally learned to drive a car!

SIDEBAR

Shirt-tail signers and their comments, some of which make no sense at all after 50 years:

What's he better at, women or flying? Or are they both the same?
—John Higson, White Plains, N.Y.

If you can handle a plane as smoothly as you can handle women, you'll make it!
—Ed Pruett, Nyack, N.Y.

Dat's my boy.
—George Schrauth, Richmond Hills, N.Y.

How's the English coming along?
—Tom Wilbor, Noroton, Conn.

Dick Payne, Cadillac, Mich.
"Chuck" Svoboda (Lawrence, Kans.)
M.L.Harvell, Hattiesburg, Miss.
"Woody" Rupp, Saginaw, Michigan

Don't bounce your landings.
—Jim Gillcrist

Glenn H.Morgan, Terry, Miss.

Warm Stone?
—Martin Manasse, New York, New York

Doug Drake, The Lover From Troy, OHIO, that is.
FROM ONE HOT ROCK TO ANOTHER

—NICK (Vagianos), N.Y., N.Y.

Happy Landings!
—Art Young, N.Y.

Keep Em Frying
—Joe Hogan

To the biggest lady killer in the unit. Lots of luck.
—"Eggie" (Facioli), Nyack, N.Y.

Nice Going
—Sam Meredith, "Scarsdale" (N.Y.)

"LONG LIONA"
—Russ Roberts

In Red No less—Congrat's
—Ed 'House'pian

TO THE GUY IN THE NEXT SACK WHO TALKS IN HIS SLEEP.
FROM KEN HORN, DAYTON, O.

John Greacen "Greek" Scarsdale, N.Y.

"Davey" Jones. Welcome to my locker. "Oatley"

From The BANZAI Kid
—Bill Gillen, Brooklyn, N.Y.

To a guy who flys as fancy as he printed up this rag!
—George Zaimes

Nice going!
—"Bud" Hower, Scranton, Pa.

CHAPTER 3

"The Schneider Exam"

by Fred "Crash" Blechman

Before you could fly Corsairs in the Navy, you had to go through flight training—and before you could get into flight training, you had to pass a physical exam. In those days the standard Navy flight physical included the "Schneider Exam."

First Time

My first recollection of the Schneider Exam was back in 1945 when in New York I first entered the Navy V-5 program—two years of college before flight training. I had no problem passing the physical, finished the college requirement, and soloed an N2S-4 Stearman in Dallas, Texas, on September 16, 1946. However, as described in Chapter 2, I left the Navy soon thereafter.

I maintained contact with a former classmate and found the flight training program had radically changed, so in the Fall of 1948 I decided to get back into Navy flight training as a NavCad. But my life had gone through a number of changes; I was two years older, and in a totally different environment.

After leaving the Navy in 1946 I left New York and entered aeronautical engineering training at Cal-Aero Technical Institute in Glendale, California, at the Grand Central Air Terminal, near Los Angeles. (Both the school and the airport are gone now. The former airport is an industrial complex.)

I was going to school under the G.I.Bill, which paid the school tuition, and provided minimum living expenses. I survived in a cubicle in a former WWII barracks at the airport, but needed some money for other expenses. So, In addition to studies during the day, in the evenings I worked in the school cafeteria washing pots and pans. Then, at about 8PM I took a bus to a local Sears warehouse and assembled steel shelving for four hours each night, catching a late bus back to the barracks.

When I took my first entrance physical at the Los Alamitos Naval Air Station—itself a long, tiring bus-ride (with transfers) from Glendale—I was not in the best of shape. And this became apparent with the results of the Schneider Exam.

The Exam

The Schneider Exam was intended to provide a measure of circulatory efficiency. It basically involved a medical technician taking applicant blood pressure and pulse readings three times—laying down, standing up, and after light exercise. The exercise consisted of stepping up and down on a stool ten times, as I recall. The results of each of the blood pressure/pulse readings were then entered into a complex chart that yielded a number, with "18" being best.

My reading came up "8." The technician was troubled. "Well, although 8 is passing, we really are looking for 11 or better for a NavCad applicant. Why don't you come back next week?"

Next week I returned, and the Schneider Exam came up with a "4"! Wow! I had gotten worse, probably a result of nervousness at having not passed the first time, and also all the hours I was studying and working. In two following weeks I returned and got a "0" and a "-4" on my Schneider Exam! Boy, was I nervous! (To this day when the blood pressure cuff is put on my arm in a doctor's office I have a psychosomatic reaction that raises the readings well beyond normal. . .)

"Hmmmm," the officer-in-charge remarked as he looked at

my medical report after the last try. "It seems you passed the first time, but have gone down hill ever since. No doubt you are too concerned about passing the Schneider.

"Tell you what. We have not reached our quota of NavCad recruits and you certainly qualify in all other respects, so we're going to change the initial Schneider reading of 8 to an 11, so you can qualify as a NavCad to go to Pensacola—but don't tell anyone. Is that okay?"

Needless to say, I was elated! Pensacola, here I come!

Party Time

In early November I received my orders to take the train to Pensacola for Pre-Flight. My buddies at school decided to have a going-away party for me the night before I was to board the train. The guys invited their local girlfriends and we gathered at a local watering hole.

I was not a "drinker" then (and am still not) and I thought I was drinking Coca-Cola all evening. I later found out that every time I turned my back, someone poured some whiskey into my glass, until eventually I apparently was drinking straight whiskey without knowing (or caring, at that point)—and I finally got sick and passed out!

I'm told I was literally carried aboard the train to Pensacola the next morning. For three or four days as the train slowly crossed the country from California to Florida, I was a mess. Sick. Weak. Miserable. And by the time we got to Pensacola I really didn't give a damn about the Navy, flight training, or anything else. I was a zombie.

On my entrance physical, with "18" being best, "11" desired, and "8" passing, I got a "14"—probably because my former nervousness was replaced by complete complacency. I just didn't care if I passed or not.

So much for the efficacy of the Schneider Exam—and perhaps the reason it was dropped from Navy flight physicals many years ago. . .

CHAPTER 4

"Instrument Flying"

by Fred "Crash" Blechman

You've just taken off in your Corsair with a supposedly thin layer of low status clouds right ahead of you. You've retracted your wheels and flaps and set your propeller rpm, mixture control, manifold pressure, and trim tabs for a nice, steady 130-knot climb. A little bumpy, but now it was just a matter of joining up on the Corsair that had just taken off ahead of you as soon as you penetrate this thin layer of clouds.

The plane ahead of you disappears into the clouds, but you continue peering through your windshield while climbing, hoping to spot him as soon as you clear this cloud layer. A few minutes pass, in this nice steady climb, when something gets your attention. The engine sounds like it's racing, and the wind past the canopy sounds like you're diving rather than climbing!

You look around and see there is no horizon; you are in the clouds. In a Corsair, with the cockpit behind the wings, when in a climb you do not see a horizon—and you certainly don't see one in the clouds! You suddenly realize you should have been flying on instruments.

A glance at your instruments is shocking! A quick scan indicates you are in a left, nose-down turn, losing altitude, at an airspeed of 220 knots. But this couldn't be. You are sure you are in a steady, wings-level climb. The instruments must be faulty.

Trust your Instruments

You were wrong, and your instruments were correct. You were in a classic "grave-yard spiral," experiencing a classic case of vertigo, and if you continued to depend on your SENSES rather than your INSTRUMENTS, you were on the way to a fatal crash!

I know because this happened to me (CHAPTER 8), and Herb Landess was killed and almost took Cookie Cleland with him (CHAPTER 41) as a result of vertigo.

Unless you've experienced vertigo in an airplane, it's hard to describe. When flying you are subject to six-degrees of motion (forward/back, up/down/, left/right) and rolling about three axes (roll, pitch, and yaw)—all simultaneously, especially in bumpy air.

Your human balance sensors—three semi-circular canals in the inner ear, ingeniously "designed" so each is on a different mutually-perpendicular axis—together with visual clues, provide your brain and muscles with the proper signals to keep you standing and walking (or running) without difficulty. But without visual guidance, the inner-ear senses become confused in the flying environment, and produce false "signals." If you ever get into instrument flying conditions, you MUST learn to IGNORE your senses and depend on your instruments!

Your Senses

You were built to walk, stand and ride on the surface of the earth. So long as you stay there and can see fixed directional landmarks, you have a fairly accurate idea of where you are is going. You don't need to look at flat surfaces, vertical superstructures, or other objects that are parallel or at right angles to you in order to keep your physical balance in this environment. Close your eyes or turn out the lights at night, and you still won't fall on your face. In the dark, or with your eyes shut, you still know which way is up and which way is down. But get into an airplane where unfamiliar forces are at work, and obscure your vision or be denied visual landmarks, and you lose both your sense of balance and your sense of direction!

As long as you remain on the ground, all your senses give you cues that are helpful in maintaining balance and in determining direction and motion. Of all these sensations, VISION is the most dominant and reliable. Many sensations are just as reliable when you fly as when you move on the ground, but others become most UNRELIABLE when you are subjected to the movements and forces of flight. Whenever you can SEE familiar horizontal or vertical references, you pay no attention to the other sensations, since vision dominates all others.

Sometimes, however, darkness or other conditions of poor visibility obscure familiar references. It is usually then, if you trust false impressions of sensations other than visual, that you can become a victim of "vertigo"—confusion about exactly what your airplane is doing.

You need not be troubled with vertigo if you learn to interpret and TRUST YOUR INSTRUMENTS. This may be called "vision through instruments." It's a matter of constantly scanning your instruments and mentally interpreting their readings to provide a mental picture of the airplane's attitude and performance.

You can't fool around with this. You must INSIST that your body obey the signals provided by the instruments, with the strict adherence to LOGIC—the "picture" provided by the instruments—instead of instinct. If the senses tell you are climbing, but the altimeter is dropping, the airspeed increasing, and the vertical speed indicator is negative, you are NOT in a climb! Even if the needle-ball is centered, you could be in a dive. If you are fortunate enough to have a gyro-horizon, that SHOWS you your aircraft's attitude.

Your aircraft is constantly operating under a simple formula: ATTITUDE + POWER = PERFORMANCE. Once you've established the airplane's attitude by visualization of the instrument readings, you provide the necessary power for the desired performance. If your nose is up, but you are losing airspeed, add power. If your nose is down, and a wing is down, reduce power, level the wings, and pull back on the stick, adding power as you approach level flight. You must force your muscles to obey your mind's visualization no matter how wrong it seems. If your senses get excited and out of control, the plane will follow them, and that can be fatal.

The Link Trainer—The Pilot Maker

(The following excerpt is from the American Airpower Heritage Museum Internet website, used with permission: http://www.avdigest.com/aahm/trmafspc.html).

The Link instrument trainer was the world's first successful flight simulator which taught basic instrument flying, radio navigation, and instrument landing to more than 500,000 military pilots, both Allied and Axis, during World War II. More than 10,000 trainers were built and sold to 35 nations including Germany, Japan, Italy and Russia. It is doubtful that any other single military training device had such wide spread use by both the Allied and the Axis powers during World War II.

The Army Air Force purchased 11 different models, and the Navy purchased 19 models. Each aviation cadet received 40-50 hours in the Link trainer. Many pilots continued to rely on the Link as a refresher course throughout their military careers. Most pilots give the Link Trainer credit for developing their instrument skills to the point of getting them out of hazardous situations later.

The Link Trainer is the nearest thing to actual flight in a World War II aircraft that can be safely provided.

Inside the Blue Box

The Link or the "blue box" is a totally self-contained trainer. All motions and electronic signals are generated from within. The Link trainer consists of a large electric turbine, which is nothing more than an oversized vacuum cleaner, that provides suction power to operate more than 20 bellows of varying sizes. Four large bellows provide the pitch and roll motions, and a ten bellow motor provides the rotation about the vertical axis. Small bellows are used to produce spins, stalls, rough air conditions and to make trainer instruments respond like aircraft instruments.

All electronic signals used for radio navigation and instrument landings are generated within the trainer—no outside radio station signals are required. Synchronous motors are used to cause the recorder on the instructor's desk to duplicate the path flown by the pilot in the trainer. The recorder depicts the flight path in two dimensions only; altitude changes made by the pilot must be marked on the chart by the instructor. Voice communication between the student and the instructor is by a type of phone, not radio transmission.

Instrument Training

Before you could fly Corsairs, you trained in the North American SNJ Advanced Trainers. Part of your training was flying on instruments. Initially, you spent time in Link trainers earning to "fly blind." You were completely enclosed as the Link trainer moved through each of its three axes (roll, pitch, and yaw) while you tried to keep the "aircraft" on a specified altitude, heading, and airspeed. Then you advanced to "flying the range," radio beams that used to be used for radio navigation. These are now extinct, replaced by much more reliable and far-simpler-to-use electronic instruments.

Once schooled in the basics of Link instrument flying, and mildly aware of how your senses could misdirect your actions, you climbed into the back seat of an SNJ, with the instructor in the front seat, for an actual instrument training flight in the air. (Usually the instructor was in the rear seat on visual flights.) After take-off you pulled a canvas cover over your head, latched it to the instrument panel, and your only visual clue to your plane's attitude and performance were the instruments!

You might be training under "full panel" or "partial panel". A full panel of instruments consisted of the gyro-horizon, directional gyro, turn indicator, balance indicator (ball), altimeter, airspeed indicator, vertical speed indicator, magnetic compass, manifold pressure gage, and tachometer. A partial panel meant the gyro-horizon and directional gyro were not used, either by caging them or (preferably) covering them up.

After a period of learning to fly straight and level, and to make "standard" turns, climbs and descents, the instructor would put the plane in an unusual attitude (you could be inverted!) and have you pull it out while under the hood! This is when you discovered that confusion in interpreting the instruments caused vertigo. . . and that in a REAL LIFE situation, you must depend on your instruments.

ABLE, BAKER. . . and CHARLIE

Various imaginary "patterns" were used in instrument training, initially practiced in the Link, then in the air. The ABLE Pattern was the simplest: Maintaining cruising speed and altitude, fly North for one minute, then make a 90-degree right turn (at the standard 3-degrees per second turn rate) and repeat this for three more legs, ending up where you started (in no-wind conditions) six minutes after you started. Even this simple two-dimensional pattern took practice under the hood.

The BAKER Pattern was a bit more elaborate, but still only two-dimensional in that there were no changes in altitude. All turns were at the standard 3-degrees-per-second rate. Maintaining cruising speed and altitude, you would fly North for two minutes, 270-degree left turn, two minutes, 450-degree right turn, two minutes, 270-degree left turn, two minutes, then final 450-degree right turn to end up where you started, but sixteen minutes later.

But the really tough one was the CHARLIE Pattern. This involved a pattern similar to the BAKER pattern, but with various changes in altitude at the standard rate of 500-feet-per-minute—sometimes even when turning—and changes in airspeed! Turns were at standard rate of 3-degrees per second. This ten-step pattern went this way, with all straight legs two minutes long:

(1) Start heading North at normal cruise speed.
(2) Two minutes straight and level.
(3) Level 270-degree left turn.
(4) Climb gaining 500 feet and accelerate to fast cruise.
(5) Level 450-degree right turn.
(6) Level one minute, lose 500 feet in second minute.
(7) Climbing 270-degree left turn, gain 500 feet.
(8) Slow cruise two minutes.
(9) Descending 450-degree right turn, losing 500 feet.
(10) End sixteen minutes from start, in slow cruise.

Whew! This was a real workout, but was GREAT training for the Fleet, where it was not uncommon for us to fly under instrument conditions.

Corsair Instruments

When I got into advanced training flying Corsairs, the instrument training was somewhat different, since this was a single-cockpit airplane. The pilot in training used a hood that slipped over his helmet and obstructed the view through the canopy, while another plane flew nearby as a safety pilot—to be sure the trainee didn't get into a dangerous flight condition, and to make sure there were no potentially interfering aircraft nearby. Typically, two planes would go up and "switch" flying instruments, communicating by radio.

We could practice flying the radio ranges that existed in those days. These consisted of four radio beams emanating from a stationary site, dividing the area into four quadrants. Alternate quadrants had a Morse Code "A" or "N," and the "beam" would be a steady signal (since it combined the dit-dah of the "A" signal with the dah-dit of the "N" signal).

A series of Morse Code letters would identify the station. Standard turning procedures were used to determine which quadrant you were in, and whether you were flying toward or away from the station. When you were in the immediate vicinity of the station, you would enter the "cone of silence" as you flew by.

Another thing we practiced (although I only recall doing this in the Fleet, while in VF-14, not in flight training) was flying GCA (Ground Controlled Approach.) Here, again, under the hood, you flew (with a safety plane alongside) a verbal approach to a runway. Ground controllers would have you on radar that defined your altitude above-or-below, and your position left-or-right, of an imaginary glide path that ended near the approach end of the runway. You flew instruments as the controller would announce your condition: "Above glide path. . . turn three degrees left. . . on glide path. . . on heading. . . below glide path. . . add power. . . turn right five degrees. . . coming up nicely. . . on glide path. . . on heading. . . etc., etc." Some would be more specific, like "Twenty feet above glide path. . . " In any case, you did the best you could to stay on heading, on glide path, and at about 100-feet (or was it 200 feet?) above the runway level, you would pop the hood and spot the runway for a visual landing. GCA did not take you to touchdown.

I always found instrument flying challenging, and a lot of fun in the Link trainers, where it was like playing with a pin ball machine in a way—trying to get all the needles, gauges, and the ball to do what you wanted them to do. In the real world, however, if you didn't trust your instruments when you lost outside visual, you were a dead man!

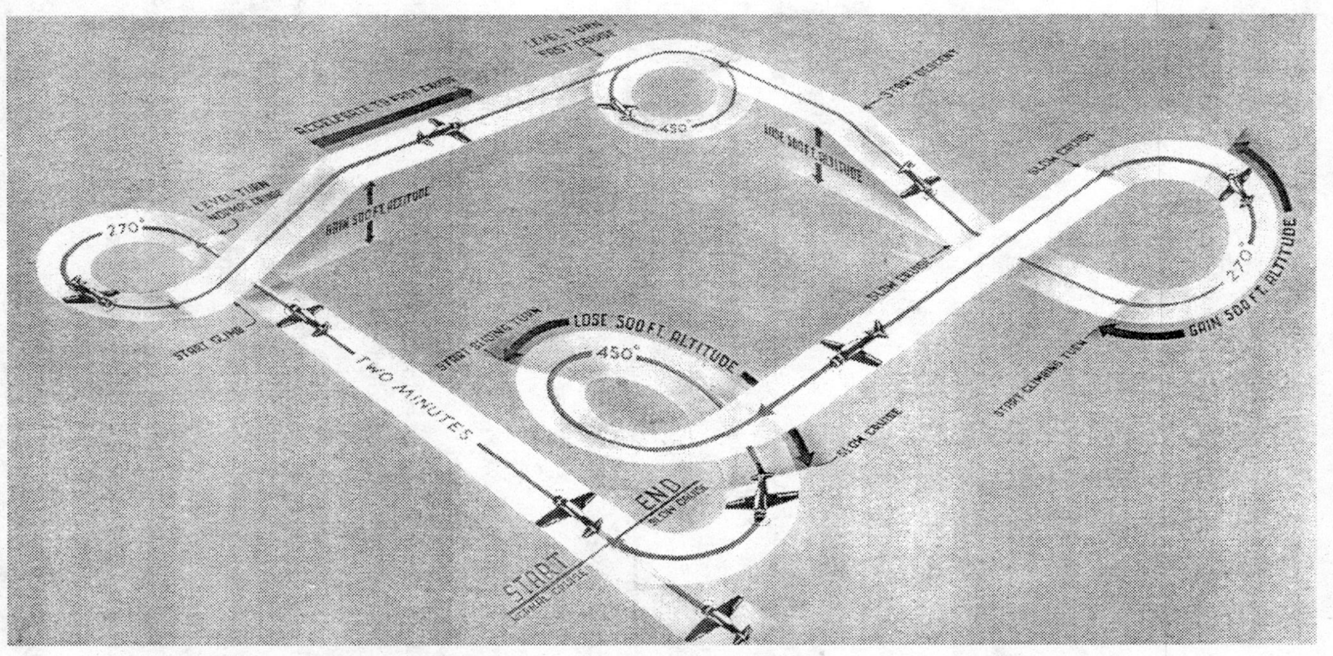

Flying the "Charlie Pattern" on instruments was a real challenge, with frequent changes in direction, altitude, and speed—all at "standard" rates. Total time: 16 minutes.

CHAPTER 5

"Terror at 100 Feet!"

by Fred "Crash" Blechman

It was May 8, 1950, and I was scheduled for my first flight in an inverted gull-wing F4U-4 Corsair, commonly known as the "Ensign Eliminator." Although I was not yet an Ensign, but just a lowly NavCad, this was still quite a jump from the SNJ trainer I had flown for over 200 hours in Navy basic flight training. I had "graduated" by making the required six arrested landings in the SNJ at Pensacola aboard the U.S.Cabot light carrier (CVL-28), and was now at Corpus Christi, Texas, for "advanced" training in Corsairs.

All our flight training up to that point had been in the two-seat 600-horsepower SNJ. We had an instructor on the dual controls for the first 20 flights before solo, and on all new types of flying—aerobatics, gunnery, formation, night flying, air combat manuevering, cross-country navigation, and so forth.

Although I had many solo flights in the SNJ, flying the single-seat 2100-horsepower F4U-4 Corsair was going to be a strictly new experience—no instructor on dual controls. I read the manual and went through a blindfold cockpit check to be sure I knew where the various controls were located, and how to operate them; some went up and down, others forward and aft, and many rotated.

Now was the time. I climbed into the cockpit of F4U-4 Serial

#81728 through the use of a wing walkway, steps, and handgrip on the right side of the airplane. Settling into the bucket seat where the parachute had been already put in place by the plane captain, I snapped the shoulder straps, seat belt, and parachute harness into the complex single-release restraint system.

Following a 19-step procedure, I started the 18-cylinder twin-row Pratt & Whitney R-2800-18W air-cooled engine and watched the giant four-bladed propeller churning in front of a nose that extended more than 15 feet ahead of the cockpit. The deep-throated harumph-harumph sound and vibration of the huge engine permeated the ground, air, and airframe. This was power!

After checking the instruments for normal readings, I gave the plane captain a "thumbs up" to remove the wheel chocks and I carefully taxied to the runup area near the runway as the engine was warming up. With the huge nose blocking out all forward vision (remember, this was a taildragger), I had to alternately turn left and right about 30 degrees, "S-turning," to see what was directly ahead.

I checked the engine oil and fuel pressures and magnetos at 2000 rpm and the supercharger at 1300 rpm, all with the propeller control set at "Take-off Rpm." Everything looked OK, so I completed the 25-step takeoff check-off list—things like making sure the prop pitch was set to full rpm, mixture full rich, flaps 20-degrees down, rudder tab 6-degrees right, aileron tab 6-degrees right wing down, elevator tab 1-degree nose up, tail wheel locked, etc. This done, I taxied to the beginning of the runway and when I got my takeoff clearance on the radio, I slowly pushed the throttle all the way forward, keeping the tail down with full back stick.

It was immediately obvious that I needed right rudder to counter the left-pulling torque of this huge engine. As I quickly speeded up, I let the nose lower to a slightly-up position and the Corsair simply flew off the ground. Now it was wheels up, milk up the flaps 10-degrees at a time, reduce power, and crank the fish-bowl canopy closed. I was in the process of doing all this at about 100 feet above the ground in a slow climb with the canopy closed, when it happened!

I must interrupt at this point to relate an earlier life experience. I have an unexplained fear of things that crawl, slither and flutter around, such as spiders, moths, lizards, snakes and creepy-crawly things that seem to pervade the hot, moist climates of Corpus Christi and Pensacola, where Navy flight training was conducted in the early 1950s.

I recall a time in my youth one night when I got into a phone booth, closed the accordian doors, and the light automatically came on. Inside that phone booth, apparently laying in wait for a wimp like me was a rather large tan, powdery moth, with long antennae. It immediately proceeded to flutter around inside that phone booth. For some reason, the thought of that thing touching me, or LANDING ON ME, threw all my alarm systems to full blast! I was out of there in a flash (even though I had not completed taking off my outer garments to reveal my true identity as Superman).

Now that I've destroyed my macho image, let me get back to the story. So there I was at 100 feet on my first flight in this 2100 horsepower fighter plane, climbing out with my canopy closed, when, all of a sudden, from the innermost black depths of the Corsair's fuselage, a big, tan, powdery moth with long antennae (obviously a direct descendant of my phone booth companion) jumped into view and started fluttering around the tiny constraints of the cockpit! I was terrified. Just me and this giant moth inches away from my tender psyche while locked in a glass bowl traveling through space at a speed well over 100 miles an hour, and only about 100 feet over trees and swamps. Yech!

I had to make an immediate decision. There was no room for both of us in that cramped cockpit. I was certainly not supposed to have an uninvited copilot. Jump or stay? Of course, how could I possibly explain bailing out because I was attacked by a moth? I quickly found an alternative. I cranked open the canopy enough for a giant sucking sound, and created a vacuum that pulled that critter up and away!!

I don't remember anything else about that flight, but I must have survived. I had 250 more flights in Corsairs in flight training and the Fleet. . . and that never happened again.

CHAPTER 6

"F4U-5 Corsair Fallacies — Some Engineering Improvements That Weren't!"

by Fred "Crash" Blechman

I've flown the F4U-4 Corsair. It was a good airplane. I've also flown the later model F4U-5—and it was no F4U-4!

I flew the F4U-4 in flight training, and when I received my "Wings of Gold" in August of 1950 I was immediately assigned to the VF-14 "Tophatters" home-based in Jacksonville, Florida, flying the F4U-5. I found the F4U-5 to be heavier and more dangerous than the F4U-4, primarily because of several "improvements" that were either dangerous or ineffective.

The F4U-5, a 1945 design modification of the F4U-4, was intended to increase the F4U-4 Corsair's overall performance, and incorporate many earlier Corsair pilots' suggestions. It featured a more powerful 2,300 horsepower engine with a fully automatic two-stage supercharger. Other "improvements" were electrical trim control, automatic cowl flaps, a gyroscopic lead-computing gunsight, and other automatic functions. These and other changes made the F4U-5 500 pounds heavier than the F4U-4.

Some improvements were worthwhile. Spring trim tabs on the elevator and rudder reduced the formerly-heavy control forces about 40%. The seat was adjustable with padded armrests that would swing down, and the metal rudder pedals could swing back to expose some padding. You would stick your legs through the rudder pedal supports, rest your arms on the armrests, and light up a cigar (yes, even a cigarette lighter was added), for relative comfort on long flights.

Electrical Trim Control

However, there were some other features about the plane that weren't so wonderful. For example, electrical trim control. Two trim tab setting switches were used. Rudder trim was accomplished with a center-off LEFT/RIGHT toggle switch. Elevator and aileron trim used a five-position center-off "joystick" switch for NOSE UP/ DOWN or WING LEFT DOWN/RIGHT DOWN.

This was easy enough to use in place of the typical mechanical rotating-knob trim controls, but you had no trim knob "feel" at all compared to the mechanical method. You had to get all the feel from the stick. When the stick required no pressure, you were trimmed.

But there was always a certain amount of inertia overrun using electrical trim control. When you released the trim tab switch the trim tab wouldn't stop right at that point. So you were constantly fiddling around, back and forth, until you got the "neutral" position.

That was bad enough, but the potentially deadly problem was switch contact welding. Every time you open and close an electrical switch there is some sort of arc. It might be too small to be noticeable, but eventually the arc can cause the contacts to wear, and unfortunately sometimes weld together! This is the same as leaving the switch ON.

Several F4U-5 pilots were lost in dive-bombing practice. When you push over into a dive and the speed increases, the airplane nose wants to come up. So you are constantly feeding in NOSE DOWN trim to relieve the increasing forward stick pressure. As these guys were feeding in down trim, they would release the switch. . . but it didn't turn off! They would get full nose down trim and go right into the ground. No time or enough strength to pull out. . . although I heard of one pilot (a big, strong guy) who put his feet up on the instrument panel and pulled back on the stick with all his strength, and made it!

We were then told, "Well, we have this little problem with the trim tabs. When you are up at 10,000 or 20,000 feet, before you start your dive, feed in NOSE DOWN tab to where you think it SHOULD be, so you don't have to touch it on the way down."

The rule was "don't touch the trim tab on the way down in a dive." We would have to hold a lot of back stick pressure before pushing over into the dive, releasing the back pressure slowly all the way down. This made for a lot of "porpoising," and a terrible gun or bombing platform. The stick was rarely neutral-force at the bottom of the dive, and then we had to fight the stick force back up to altitude as our speed decreased in the zoom. Needless to say, I did not like those early electrical trim tab switches!

Automatic Cowl Flaps

The F4U-5 was the first Corsair with automatic cowl flaps. The cowl flaps were tied in with the engine temperature and would open to cool the engine when necessary. This was fine unless you were in close formation and the cowl flaps suddenly opened, adding drag, dropping you behind. Worse was when you were flying on somebody's wing and the cowl flaps were partly open and decided to close! Now you would have to reduce throttle quickly to keep from overrunning your lead!

Fortunately, the cowl flaps did not HAVE to be on automatic in the air, and could be controlled with an OPEN/CLOSE toggle switch. This, of course, meant watching the engine cylinder head temperature. So, except on formation flights, we used automatic cowl flaps.

There was another automatic cowl flap switch on the airplane, located on the left landing gear scissors. Whenever the wheel strut was depressed because of the weight of the aircraft, the automatic cowl flaps would open fully for adequate engine cooling when the airplane was on the ground. That was a nice feature, since if you didn't open your cowl flaps when you were on the ground, you would burn up your engine taxiing back to the ramp!

However, because of the landing gear switch, which ALWAYS seemed to work, it was easy to overlook closed cowl flaps while taxiing. If the switch didn't work (which rarely, but sometimes happened), the cowl flaps didn't open, and the engine burned up.

Automatic Power Unit

Another well-meant improvement to the F4U-5 was the automatic power unit (APU), an "automatic blower control." On the F4U-4 and earlier Corsairs a manually-operated supercharger control would allow you to get an extra ten inches of manifold pressure on take-off, and essentially maintain power at altitude. On the F4U-5 this was all done automatically, WITH NO MANUAL OVERRIDE!

The APU monitored manifold pressure, air temperature, atmospheric pressure, prop rpm, and a number of other parameters. These readings were fed into the APU's computer, which decided when it needed the extra power. We could not make formation takeoffs in the F4U-5 because nobody knew when his APU would cut in. When making a carrier deck-run takeoff, we could get off the end of the deck and be halfway down to the water before the APU would decide to cut in. On catapult shots, we didn't give our "Ready" salute to the Launch Officer until the engine manifold pressure gauge showed the extra ten inches!

If we were in formation, we would have to split up at about 19,000 feet going up, since the APU would activate our second-stage supercharger somewhere between 20,000 and 22,000 feet. We could not control when our blower would cut in, so we had to separate a good distance apart. When all the blowers had engaged, each pilot would report in and then rejoin in formation. When descending from altitude we did the reverse, breaking formation at 23,000 feet and reforming at 19,000 feet.

Gyroscopic Gun Sight

The F4U-5 also introduced the Mark 8 gyroscopic lead-computing gun sight. This gyro gun sight was created to simplify the lead "deflection" shot. When attacking a target that is moving across your flight path, you have to "lead" the target so that the bullets leaving your guns will follow a path that intercepts the moving target.

The Mark 8 gun sight projected an image onto the windscreen. The image had six diamond-shaped "pips" that formed an imaginary circle. Twisting the throttle handle's knob clockwise made the circle smaller, and twisting it counter-clockwise made the imaginary circle larger. The object was to encircle the wingspan of the target within the imaginary circle. As you flew closer to the target and it got bigger, the idea was to turn the throttle handle slowly and steadily counter-clockwise to keep the target tightly encircled.

You were also flying the airplane, moving the throttle forward and backward as necessary, as well as operating the rudders while moving the stick all over the place. After all, you were trying to home in on the target. At the same time you were supposed to be feeding in this information to the computer by turning the knob, telling it how close you were to the target aircraft and your rate of closure. With this information, the Mark 8 "calculated" the deflection angle needed and MOVED THE CIRCULAR IMAGE TO A NEW POSITION on the windscreen!

So, as you are flying the aircraft, this thing is floating around the screen. You are constantly adjusting and trying to hold your target in the center. If you jerked the throttle a little bit, the whole thing would suddenly move up, down or to the side. You would try and be as smooth as you could, but there is no way a human is going to be smooth enough to do this!

Actually, the idea was good, but putting a human in the control loop was not working. In the F-86 Sabre jets used in Korea they let a small radar set in the nose determine the range and closure rate to the target. All the pilot had to do was keep the target inside the slowly floating imaginary circle, and fire when within range. This was highly successful.

Other "Improvements"

There were other changes that were surprising. Why did they go from a retractable tailhook to one that had to be manually raised after a carrier landing? Why did they put the plotting board into the instrument panel with such a weak clip? Sometimes it would pop back into the pilot's chest on a carrier catapult shot. And those nice soft-backed rudder pedals would sometimes snap back on a catapult shot and the front metal part would smash onto the pilot's ankle.

All in all, as you might gather, I preferred the F4U-4. . . and so did the Navy. During the Korean action, only some F4U-5N night fighters, and some F4U-5P photo reconnaissance versions were used. Plain-vanilla F4U-5s were pulled back from the front lines and replaced with. . . you guessed it. . . F4U-4s!

CHAPTER 7

"The Great Gum Mystery"

by Fred "Crash" Blechman

I wonder if something like this has ever happened to you when you were flying?

This mysterious incident took place in 1951, while I was flying with VF-14 at Cecil Field in Florida. I was flying the last F4U-5 Corsair in a flight of eight on a practice dive-bombing mission.

As usual, I was chewing gum—Dentyne, my favorite. As I recall, up to that time I just about always chewed gum when I was flying. This, apparently, was not uncommon among pilots, and it has become well known that the famous WWII and test pilot, Chuck Yeager, often could be found bumming a stick of Beeman's Chewing Gum when he ran out.

We climbed to about 15,000 feet above the practice area at an outlying field, preparing to dive-bomb our target with small smoke bombs. The scenario was to be in a loose right echelon above and to the right of the target area. We would drop the wheels to act as dive brakes and preset the electric elevator tab for a dive. (NOTE: We didn't dare touch the nose-down tab switch in an F4U-5 during a dive, since some switches had been known to weld shut and put the tab in FULL nose down, from which it was virtually impossible to recover when in a dive!)

We would then peel off to the left for about a 180-degree change of heading as we spotted the target, dropped our nose and set up about a 60-degree dive for the bulls-eye markings. Another

Corsair, orbiting the target at about 1500 feet was acting as a spotter. After we dropped our bomb and pulled up in a 6G pullout, straining us even in our G-suits, we could check our accuracy by looking directly behind us with three rear-view mirrors mounted at the very top of the cockpit canopy.

It was beautiful to watch as each plane peeled off in five second intervals and aimed back and down—just like the WWII flying movies. As my turn to peel off approached, in my eagerness I began to chomp on the gum more furiously—when suddenly it DISSOLVED! It just simply turned into a sand-like texture and slipped down my throat in a granular stream! Gulp! (Got milk?)

I was so distracted I missed the target by more than my usual distance. . . and thereafter stopped chewing gum while flying. Is it possible that I plain pulverized the gum by chewing it at such a pace? Does it have something to do with altitude? Has it ever happened to you? Any explanation?

CHAPTER 8

"A Flight to Remember"

by Fred "Crash" Blechman

They say a cat has nine lives. I must be part feline, or I would not have survived my tour of duty with Fighter Squadron Fourteen (VF-14) in the early 1950's. In particular, February 13, 1951 would have ended it all for me.

I had completed 21 months of flight training when I earned my Naval Aviator wings and Ensign bars on August 23, 1950. In early September I reported to VF-14, then stationed at Cecil Field, outside of Jacksonville, Florida. Although, like all red-blooded American Heroes, I had requested West Coast duty (since the Korean fracas had just begun), the Navy, in their infinite wisdom, sent me instead to the East Coast of Florida.

The Commanding Officer was Lieutenant Commander Robert C.Coats, now a retired Captain living in Jacksonville. Skipper Coats took me under his wing, since I was the junior pilot in the squadron—and remained so for over a year, until Ensign Gene Hendrix (now a retired Lieutenant in Fort Walton Beach, Florida) joined the squadron. Most of the 24 pilots in VF-14 were seasoned naval aviators, several having been called back to active flying duty from the Navy Reserve after serving in World War II.

The squadron aircraft was the latest model Chance-Vought Corsair, the F4U-5. This was a jazzed-up version of the F4U-4 inverted gull-wing fighter I had flown in advanced training, and in which I had made six carrier landings before qualifying for my wings. The fact that the Corsair had earned the nickname of "Ensign Eliminator" was not exactly consoling.

My first F4U-5 flight was on September 20. By this time I had become acquainted with some of the techno-whiz additions that had been designed into the F4U-5. I mean, do you really need a cigar lighter, padded arm and leg rests, electric trim tabs, computer-controlled engine boost, gyro gunsight, automatic cowl flaps, and such? Aside from adding considerable weight and complexity to this aircraft, some of these "improvements" turned out to be real potential killers—as you'll see—and some were just plain annoying.

The squadron was scheduled for a Mediterranean cruise in early 1951 aboard the USS Wright (CVL-49), a light carrier. Since the squadron had only recently received the new F4U-5s, Skipper Coats' main task was to get us all carrier-qualified in the F4U-5. This meant many FCLP (Field Carrier Landing Practice) flights to a nearby field that was marked out like a carrier deck, and using an LSO (Landing Signal Officer) to bring us in for touch-and-gos.

Although the "deck" wasn't moving, it wasn't 50 feet above the water, either—it was at ground level. This meant flying very low and very slow just above stalling speed, usually over marshy ground with bumpy air. In many ways, FCLP is more challenging than actual carrier landings. My logbook shows exactly 109 FCLP landings during the three months preceding that cruise.

The squadron flew to Norfolk in late December and carrier qualified, both individually and as a squadron, by each pilot making six F4U-5 landings aboard the newly-commissioned USS Oriskany (CV-34), which was on its shakedown cruise off the Virginia coast.

Our aircraft were loaded by crane aboard the Wright, and we left port on January 10, 1951, headed for Gibraltar. We only flew three days on the ten day trip to the Mediterranean. After Gibraltar, we made the ports of Oran in Algeria, Augusta in Sicily, and Naples and Palermo in Italy. On Sunday, February 11, we left Palermo and headed for Suda Bay on the Greek island of Crete. We flew on the 12th and 13th. The flight on the 12th was uneventful. The flight on the 13th was almost my last one.

The weather was bad on the 13th as we were cruising through the Ionian Sea. Our intended mission was to practice dive bombing a small island target off the west coast of Greece. The Greek government was very cooperative, especially since our next port was on Crete.

Meteorology assured Primary Fly that the weather over the target was perfect, and that the overcast above us was only a few thousand feet thick. Looking from the flight deck, it seemed like the world was coming to an end. The low stratus clouds seemed only a few hundred feet off the deck, with fingers of clouds reaching down to touch the water. Scud clouds slid by, and there were obvious rain showers in several directions.

Eight planes were scheduled for this flight—two divisions of four planes each. The flight leader was Lieutenant Commander Felix Craddock, the squadron Executive Officer. He was to lead the first division, and Lieutenant Junior Grade A.G Wellons (now a retired Captain in Jacksonville) was to lead the second division, with Ensign Jim Morin (now a retired Rear Admiral in McLean, Virginia) as his wingman. I was, as usual, tail-end Charlie—the last plane in the last two-plane section of the last division. I was to fly the wing of Lieutenant Junior Grade "Doc" Mossburg (where are you now, Doc?)

The flight deck was heavily spotted with aircraft, so we were each launched in turn from the Wright's two hydraulic catapults. This was before the days of the slant-deck carriers with four powerful steam catapults and 1100 foot decks. The entire deck of the Saipan-Class Wright was only 600 feet long.

When my turn came, I unfolded my wings as I taxied forward, following the flight deck crew's hand signals. When I reached the catapult area, I dropped full flaps and checked my wing-lock indicators. After tightening my seat belt and shoulder harness I set the trim tabs slightly nose-up, right wing down, and right rudder, to counter the left-turning torque of the giant four-bladed prop at full power. The propeller control was set to maximum RPM as I checked the engine instruments. I made sure my G-suit was plugged in, especially important since we were headed out for a dive bombing mission. I also made sure my oxygen flow indicator was working.

While I was doing this, a deck crewman hooked up the catapult shuttle to pull against the "bridle", a steel cable that attached to a hook under each wing. At the rear of the plane another crewman attached a small steel ring to the deck and a short cable that grabbed a hook behind my tailwheel. This "holdback ring" was precision manufactured to restrain the Corsair at full engine power, but break apart when the catapult shuttle added its force. These sometimes parted prematurely, leading to a "cold shot"—off the bow with too little speed to stay airborne. Splash!

Also, sometimes the bridle would snap partway down the deck during the catapult shot, throwing the aircraft off the side of the deck into the water. Nice things to contemplate while waiting for the signal to pour on the coal.

The wheel brakes, applied by pressing the top of the rudder pedals, were no longer required once hooked up to the catapult and restrained by the holdback ring. In fact, you certainly don't want to apply the brakes while being catapulted! Therefore, the standard procedure was to drop your feet off the rudder pedals so your heels were on the deck, and your toes at the base of the pedals. I did this.

Finally, the Catapult Officer gave me the wind-up hand signal to go to full throttle. As I advanced the throttle to the stop, I checked the engine instruments. It took only a few seconds for confirmation that the engine had powered up and sounded right, and that the instruments read normally. The Corsair's over 2000 horsepower was straining against the small holdback ring, waiting for the force of the catapult shuttle pulling against the bridle to snap the ring and set it free.

I saluted the Catapult Officer to let him know I was ready, and threw my head back against the cockpit headrest in anticipation of the launch jolt. I glanced out of the corner of my eye to see when the Catapult Officer dropped his arm. That was the signal for the Catapult Operator to press a big red button to fire the catapult. The Catapult Officer watched the gentle pitching motion of the ship to make sure he was not going to fire me off with the ship's bow aimed at the water, and at the right moment he dropped his arm. Wham!

When the catapult fired, the holdback ring ruptured, just as it was designed to do, and the aircraft lurched forward, completely out of my control for about three seconds. I immediately felt a sharp pain above my left ankle, but was too occupied to find out why.

Off the end of the deck at barely flying speed, I was headed for the water. The damn APU (automatic power unit), an F4U-5 innovation, was supposed to cut in and add about 10 inches of manifold pressure—enough power to remain airborne. It had not engaged.

The electro-mechanical APU used various sensors to look at the air pressure and temperature, engine RPM and manifold pressure, prop pitch, and who-knows-what-else to determine when to add the extra power boost. This had replaced the manually operated supercharger in the -4 Corsair, and was completely automatic with no manual over-ride. A-r-g-g-h!

I quickly retracted my wheels. It was bad enough to hit the water in a radial-engine aircraft (which usually dug in and flipped the aircraft), but going in wheels down was even more likely to flip her. I left the flaps full down and milked the nose up carefully. The flight deck was only a little over fifty feet above the water, but the slight drop had increased my airspeed. Just then, about twenty feet above the surface of the water, the APU cut in. Whew!

I put the nose in a climb, slowly raised the flaps, and adjusted the power settings and trim tabs to neutralize the controls. I looked ahead to see Doc Mossburg's Corsair disappear into the low clouds. No other plane was in sight, since they had all started up through the overcast.

My ankle was still hurting. Looking down I found the cause. The heavy metal rudder pedals in the -5 Corsair were designed with soft padding on their back, but were normally held upright with springs. A pilot on a long flight could pull the top of the rudder pedals toward him and stick his legs through the pedal supports to rest his thighs. The left pedal apparently had a busted spring. It had flipped back, from the force of the launch, and the metal part had slammed against the top of my foot. Not disabling, but certainly distracting.

Intent on catching up with Doc for a snappy rendezvous on his wing, I kept peering through the windshield for sight of him. I didn't realize that I had also flown into this gray overcast and should have been flying on instruments. Somehow the overcast appeared to be above me, and it seemed like I could see for some distance ahead. I was expecting to spot Doc's plane any second. I felt like I was in a perfectly normal climb, and in a Corsair, with a huge nose projecting over fifteen feet in front of you, you don't see the horizon in a climb if you're looking straight ahead.

I had been flying in this "steady climb" for several minutes when I suddenly realized the engine sounded strange—like it was running faster than normal. Also, the wind noise in the cockpit sounded like I was flying much faster than in a climb. I glanced at the instruments. If my hair could have stood up under my hardhat helmet, it would have!

My artificial horizon showed that I was in a nose-down left turn. My altimeter was winding down furiously. My rate of climb indicator was way in the negative. My turn and bank indicator was almost pegged to the left, with the ball in the center—a nicely balanced turn. My airspeed indicator showed over 220 knots; 130 was normal climbing speed. The tachometer showed the engine RPM at over 2900, with 2550 normal for a climb. No question about it—I was in the well-known "graveyard spiral", a classical case of vertigo.

If I had hesitated much longer, I wouldn't be writing this today, almost 48 years later. It was difficult to ignore my conventional senses and suddenly have to shift mentally into believing a bunch of instruments that seemed to be lying to me. I had heard a lot about vertigo in training, and had even experienced mild cases in Link Trainers, but never like this.

I knew that if I followed the instinctive action of pulling up the nose I would only tighten the turn. The first order of business was to scan the instruments and level the wings. I was at about 2000 feet heading down at 4000 feet per minute, so I had less than 30 seconds. . . . . . I threw the stick to the right and added some right rudder to level the wings, reduced the throttle, then pulled back on the stick as I watched the altimeter and rate of climb indicator. I leveled off at 600 feet altitude without ever breaking out of the clouds!

I advanced power and went into a climb—this time on instruments. When I broke through the overcast at about 4000 feet—it was much thicker than we had been told—there were the other seven planes, all joined up in a gentle left rendezvous turn, wondering where I was.

This was not my only "close call." On other flights there were other incidents. I had a loose oxygen mask and almost flew off over the horizon in a 30,000 foot euphoria. I popped an oil line pulling out of a dive bombing run and had to make a carrier landing with my windshield covered with oil. On one flight my wheels wouldn't come down, and they had to be blown down with an emergency CO2 bottle. Once my automatic cowl flaps didn't open on landing and I almost burned out the engine. Then there was that night landing where my brakes seemed to lock. . . .

VF-14 F4U-5s, as the fourth squadron in Air Group One, had a white "T" on the tail. Bright yellow was used on the top of the tail, the wing tips, the prop hub, and (not shown here) the tail cone. (Illustration by noted artist, Jim Laurier—(603) 357-2051).

My division of VF-14 "Tophatter" pilots during 1951 Mediterranean cruise on the USS Wright (CVL-49). Left to right: ENS Merle A. Rice (deceased): LCDR Felix B. Craddock, XO: LTJG "Doc" Mossburg (deceased); ENS Fred Blechman.

CHAPTER 9

"Flying the Hayrake"

by Fred "Crash" Blechman

For many years, in the early days of flying, pilots "flew the airways," which consisted of low-frequency radio signals sent out into four quadrants roughly 90-degrees apart. Adjoining quadrants each had a Morse code A (dot-dash) or N (dash-dot) signal, and where the quadrants joined the A and N signals combined to create a solid tone. The pilot simply found the solid tone—the "beam"—and then flew inbound to the "cone of silence," where the beams converged at a specified fixed location from the airport.

But how could a pilot during World War II or the Korean "police action" find his moving carrier from many miles away? Of course, if the plane had a working radar, it was relatively straightforward. But most carrier planes at that time did not have radar, and those that did frequently had their radar fail.

The solution was ZB/YE, also known by some as "The Hayrake" because of the shape of the transmitting antenna on the ship. As I recall, ZB was the receiver in the airplane, and YE was the low-frequency over-the-horizon ship's transmitter.

Harold L.Buell, in his excellent book, "Dauntless Helldivers" (Orion books), described it this way: "Each U.S. carrier had an electronic marvel on board to help the scout teams return safely. Known as the YE or ZB, it was a rotating radio signal-sending device mounted high on the carrier mainmast above the island. This device broadcast a different letter of the (Morse code) alphabet every thirty degrees as it rotated. These letters could be picked up by a receiver in a plane as it moved into the radio signal range of the carrier; from the letter heard the pilot could determine a bearing to the ship. In the corner of each pilot's (plotting) board was a compass rose. When preparing flight data before starting a search, the pilot filled in each slice of the pie with letters designated for the YE that day."

I recently found an Air Operation Memorandum from the U.S.S. Fanshaw Bay (CVE-70) dated 29 July, 1945. Among other things, it showed the YE Frequency as 672 kilocycles, with sector letters U G D K S W R M L F A N.

When I was flying F4U-5 Corsairs from carriers in the Sixth Fleet in the early 1950s, we made several cruises around the Caribbean and the Mediterranean Sea. We frequently flew over 200 miles from the carrier, with nothing but water between us, and it was certainly a welcome sound (especially in bad weather) to hear the Hayrake signal coming in. The Morse code letter received would tell us which of the twelve quadrants we were in, and we simply flew the reciprocal heading until we spotted the ship visually. If the letter received changed, we knew we had drifted from the proper heading, and adjusted accordingly. It was a great system.

When on maneuvers, the code was changed daily—and sometimes several times a day. Normally, we just used a "standard" code starting at north and moving clockwise every thirty degrees. The letters were DWRKANUGMLFS. We committed this to memory with these mnemonics: "Did Willy Really Kill A Nasty Ugly German Man Last Friday Saturday." I remember it to this day!

But Owen Dykema (CAPT USNR Ret.) remembers it differently. He flew 47 Korean combat missions in Corsairs from the carrier Princeton. In his new self-published book "Letters From the Bird-Barge," Owen recalls a different set of mnemonics for the same default code: "Did Willy Run Kate Around Naked Until George Made Love For Sheckles" Those Korean guys!

The Hayrake is now a part of history, replaced by several advances in electronic navigation aids, with GPS (Global Positioning System) today pinpointing your position in real time, and ship's radar able to vector you in from over the horizon. But those of us who flew the Hayrake (although many simply referred to it as YE or ZB) have fond memories of a simple system that worked—unless the ship's Hayrake antenna was damaged in battle, or your receiver quit!

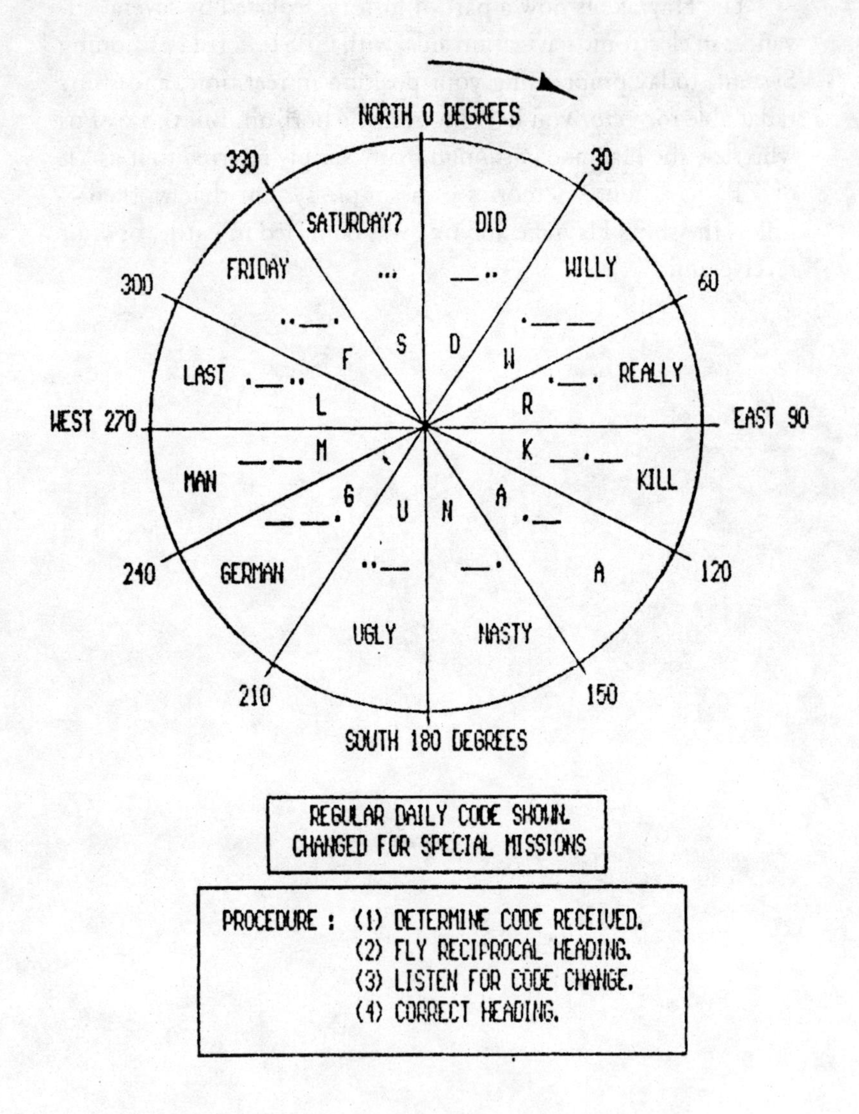

"Hayrake" default code. When in combat or maneuvers, the code changed daily. Allowed many lost Navy pilots to find their ship.

CHAPTER 10

"Over the Rainbow"

by Fred "Crash" Blechman

It was planned to be a pretty simple hop. As it turned out, I almost flew "over the rainbow" to the Land of OZ!

LCDR Felix Craddock, the VF-14 Executive Officer (XO) of our squadron, VF-14, and I were to be catapulted in our F4U-5 Corsairs from the light carrier USS Wright (CVL-49) as it was steaming in the Mediterranean Sea. I was to be Felix's wingman.

We were to fly north for about 150 miles, climbing to about 35,000 feet. We were then to turn back toward the carrier. The object was to test the ability of the ship's radar to spotting us coming inbound, and to determine our bearing from the ship, our altitude, and our groundspeed. Good practice for the ship's radar crew, and a big help in calibrating their system.

This was the Spring of 1951, and no armament was loaded, since it was peacetime in the Med (although the Korean Action was in full swing in Asia). No special equipment was needed beyond our normal helmets, oxygen masks, and liferafts. We wore our G-suits as standard gear.

The radios were checked out before we mounted our steeds—side-number 403 for the XO, and I was in #405. We climbed up and into the cockpits using the various toe-holds and hand-holds provided in the airframe for that purpose. (This took strength and the limber body of a young man, so I doubt if I could accomplish that task today at 71!). Fortunately, we weren't wearing our chutes at the time; the Plane Captains had already placed them in our cockpits.

Plugging In

As soon as we were seated, we began strapping ourselves in. An adjustable safety belt and shoulder straps both fit into a quick-release safety buckle. The shoulder straps were attached with spring tension to an inertia reel behind the seat, allowing you to lean forward yet be protected from sudden maneuvers or a crash landing when the inertia reel would lock the straps. You could also lock the shoulder straps independent of the inertia reel, which I normally did unless I had to lean forward to reach a control or switch.

I plugged my microphone/headphone connector into the radio jack located on the bulkhead behind my right shoulder. The pigtail of the G-suit was plugged into a quick-disconnect fitting located on the aft end of the left-hand control shelf.

Whenever we pulled 2Gs or more in a maneuver—even a tight turn—a valve would open and force air into the bladders in the legs, thighs, and abdomen of the G-suit. This would tend to prevent blood from flowing out of the head into the lower body, and would allow us to pull an extra G or two without graying or blacking out.

Next, I plugged the oxygen mask into a tube at the bottom of the seat, put the mask to my face, adjusted the straps for a snug fit, and breathed deeply to check that the two oxygen flow indicators were blinking to indicate proper operation.

The F4U-5 had a diluter-demand oxygen system with an air-valve lever on the left-hand control shelf that allowed the pilot to select NORMAL OXYGEN or 100% OXYGEN. I selected NORMAL for a maximum duration flight. This conserved oxygen by producing a proper proportion of oxygen that automatically increased until at 30,000 feet it was 100% oxygen. (The 100% OXYGEN selection was used on night flights for better vision, and were usually shorter flights.) Normally, we didn't use oxygen at all below 8,000 feet (5,000 feet at night), but since on this flight we were going to high altitude, I left the mask on.

The Launch

We were parked forward of the other aircraft on this relatively short deck, so we would both be catapulted—not enough room for a running deck launch. We each started up our engine, unfolded our wings, then each checked to see that our wing-lock indicator (a small red metal tab that extended above the wing) pulled down into the wing to assure positive wing-lock. We then taxied forward—Felix to the port catapult, and I to the starboard cat.

Following the standard procedure, the deck crew attached a steel bridle to hooks on the underside of each wing, with the front of the bridle looped around the catapult shuttle. At the rear of the plane, a small "holdback ring," designed to break under the forward thrust of the catapult shuttle (but NOT to break UNTIL then, although the aircraft engine was at full power!), was connected between the tailwheel and the deck.

Felix was given the signal to apply full power. When he was satisfied with his engine instrument readings, he gave a snappy right-hand salute to the Launch Officer. When the Launch Officer was satisfied with the engine sound and the position of the forward deck related to the sea (since sometimes the deck would pitch forward in a rolling sea), he gave the launch signal. The catapult shuttle strained forward, finally snapping the holdback ring, and Felix's Corsair shot forward and off the deck with a smooth launch.

The same hookup and procedure got me off the deck and into the air about 15 seconds behind Felix, who was climbing in a gentle left-hand turn so I could rendezvous with him by turning inside his radius, joining up, and then slipping under him to his starboard side, slightly below and behind him.

Formation Flying

The whole process of formation flying—from rendezvous to holding position—is a learned technique. The rendezvous is accomplished by putting the lead plane in a position on your windscreen—and keeping it there. As long as the angle stays the same, you are on a line of interception. As you begin to close in, you must adjust your rate of closure so when you get close to the lead plane you are flying a similar path and speed to his, and don't need to make any drastic control changes to stay in position.

Flying close formation is one of the most picturesque and challenging experiences of military aviation, since civilian pilots rarely fly in close formation. Watching the plane merely feet away from you moving slightly as you make tiny—and constant—adjustments to stick, rudder and throttle to stay in position takes full attention and practice.

In close formation you don't DARE look at your instruments or anything else but the plane you're formed on. The lead plane must be smooth and not make any unexpected moves, especially into the wingman. And in the clouds or at night, with no outside visual clues, you can't be concerned about whether you are right-side up or upside-down—you follow your leader. (Which can also be deadly, as you'll read about in "Cookie" Cleland's story, Chapter 41.)

The Rainbow

So, off we were, headed north over the boundless horizon, with nothing but the shimmering sea in every direction. The sky was cloudless—but ahead there were clouds above the horizon. We climbed at about 1000 feet per minute at 160 knots indicated airspeed, reaching our assigned altitude of 35,000 feet and continuing outbound.

At this point, Felix passed the lead on to me, and he flew off my right (starboard) wing. As we flew north, apparently a storm had taken place below and ahead to the left of us, and the placement of the sun and our position related to the mist in the air below created a beautiful rainbow extending from sea to sea. I was fascinated by this complete color arch, with invisible ultraviolet becoming visible blue at the inside of the arch, then through all the other visible colors and shades to red and then into invisible infra-red at the outer limits of the arch.

On we flew. I was feeling wonderful! What could be better than flying a Corsair at 35,000 feet and viewing this spectacle of nature. My radio was buzzing, it seemed. Strange voices. Shouting. It made no sense to me. Suddenly, I perceived that Felix had gone over to my left side (interfering with my view of the rainbow!) He was making strange gestures, then he pulled ahead and was flying zig-zags. I thought, "I guess Felix is bored, and is just having some fun." More radio gibberish. . .

No Blinking!

Was it training or intuition that made me think to glance at the oxygen flow indicator on the forward left console? Why wasn't it blinking? Must be bad, since I was feeling fine. But I knew there was a second oxygen indicator further back on the left console, near the trim tab controls. It was also not blinking! Hmmmm. I knew that up to 41,000 feet the indicator should blink when oxygen is being drawn. Could BOTH indicators be bad. OR MAYBE I WASN'T GETTING OXYGEN!

I pushed the mask up against my face, took a couple of deep breaths, and the radio began making sense. Felix was having a fit on the radio. "405! 405! Where are you going? Let me have the lead. I have it!" We were way beyond the 150-mile turnaround, and headed for Italy—without clearance during cold-war tensions! I was heading us to the Land of OZ. . .

I got on Felix's wing, and he turned us back toward the ship. Since I had flown us over 200 miles from the ship, it took us over an hour to get back, and after we landed aboard, I got HELL from the XO! Deservedly! I should have been checking my oxygen indicator regularly, instead of watching a rainbow.

My equipment had been working properly, but my oxygen mask straps were not tight enough, and with a demand oxygen system, it required the vacuum of your breathing in to open the oxygen valve—so I was getting little if any oxygen. Classic case of anoxia—feeling great, but with severely reduced faculties.

I believe that my former experience in an altitude chamber during advanced flight training may have contributed to my recognizing—even on a subconscious level—that I should check my oxygen indicator. I recall that a group of us were herded into a sealed chamber with two instructors and told to put on oxygen masks. The chamber pressure was slowly reduced to simulate an altitude of about 25,000 feet, and we were asked to remove our masks for about 30 seconds, then write our name with a pencil and pad provided to each of us.

The instructors left their masks on, watching and observing the rest of us. I felt fine. Good, in fact, and wrote my name with no problem. I wondered why some of the other guys were laughing, having fun—and found I was, too. A couple of the guys seemed to go to sleep, and an instructor quickly went over and applied the oxygen mask. Then we were all told to put our masks back on (Why? I was feeling great.)

When the chamber got back down to normal sea-level pressure, we got out, and got to look at our signatures. What a mess!!! They were all unrecognizable, like chicken scratchings. The point was made. With anoxia, you feel great until you pass out! I'll never know how close I got to passing out on that flight "over the rainbow. . . "

CHAPTER 11

"Don't Ever Tell. . . "

by Fred "Crash" Blechman

"Ensign, don't ever tell anyone about this! Do you understand?"

"Yes, sir!" I replied, realizing the embarrassment it would cause this senior officer. So I didn't tell—until now.

It was in August, 1951, that this senior officer—I'll call him "Harry," though that is not his name—asked me to accompany him as his wingman on a cross-country flight to Norfolk Naval Air Station from our base at Cecil Field, near Jacksonville. He had some sort of business there, and it was general practice for planes to fly in pairs in case one had engine, radio, or other troubles.

Starting the Engine

We climbed into our VF-14 F4U-5 Corsairs, buckled up, connected the microphone, headphones, G-suit, and oxygen masks, and went through the normal engine starting procedure. Lest you think this was as simple as starting your car, this was our starting procedure, right out of the F4U-5 Pilot's Handbook:

a. Ignition switch—OFF.
b. Cowl flap switch—AUTOMATIC (verify cowl flaps full open).
c. Oil cooler door switch—AUTOMATIC.
d. Intercooler flap switch—AUTOMATIC.
e. Master water injection switch—OFF.

f. Mixture control—IDLE CUT-OFF.
g. Throttle—Set to give 800-1000 rpm after engine start.
h. Propeller control—full INCREASE (low pitch).
i. Auxiliary (booster) fuel pump switch—OFF.
j. Transfer pump switch—OFF.
k. Oil dilution switch—OFF.
l. Power control switch—OFF. Connect external power source and turn propeller with the starter through four revolutions (16 blades) to clear the engine.
m. Fuel selector—ON
n. Auxiliary (booster) fuel pump switch—LOW (check that fuel pressure is at least 10 psi).
o. Ignition switch—BOTH.
p. Engage starter and prime engine.
q. When engine begins to fire regularly on prime, move mixture control SLOWLY to RICH. Do not pump or move throttle abruptly. . . it is IMPORTANT to keep the engine firing regularly (at least 350-400 rpm) by continuing to use the priming switch. If for any reason the engine stops firing, move the mixture control to IDLE CUT-OFF immediately, and continue cranking and priming until engine starts.

Two more pages in the Pilot's Handbook cover FAILURE TO START ON THE FIRST ATTEMPT, IMPROPER PRIMING, ENGINE FIRE DURING STARTING, WARM-UP, GROUND TEST, ENGINE CHECK, MAGNETO CHECK, PROPELLER GOVERNOR CHECK, IDLE MIXTURE CHECK, ELECTRICAL CHECK WITH ENGINE RUNNING, HYDRAULIC CHECK, RADIO CHECK, and WING SPREADING.

Some of these tests were performed before leaving the chocks, others after taxiing out to the runway with constant S-turning to be able to clear the path ahead. (With the long nose, we had no forward vision in this tail-wheel aircraft.)

Takeoff

Finally, it was time to take off, which we had to do individually rather than as a two-plane formation. Why? The APU (Automatic Power Unit) used in the F4U-5 cut in automatically based on various sensor inputs, and provided about 10 inches of additional manifold pressure. This gave considerably greater thrust, but had NO MANUAL CONTROL. Therefore, when two planes were taking off next to each other, the wingman could unexpectedly surge ahead—or drop way behind as the lead plane's APU cut in. So, to avoid the danger and sloppy appearance of a formation take-off, we made individual takeoffs.

The takeoff check-off list made sure the shoulder harness and safety belt were locked, the canopy was locked open, the tail wheel was locked, the rudder tab was set six degrees nose-right, the elevator tab one degree nose-up, the chart board locked into the instrument panel, and various switches and controls were properly set.

Finally, since we had plenty of runway, we made our takeoff with no flaps. (Aboard ship we would use the full 50-degree setting for the flaps for the shortest deck run—and with increased elevator tab setting.) The takeoff itself was pretty simple. You released the wheel brakes, advanced the throttle slowly to about 44 inches of manifold pressure, holding the stick back to keep the tail down, and applying considerable right rudder to counteract the left-pull of the engine torque. Then as you increased forward speed you pushed the throttle to the stop, slowly letting the tail rise as forward speed increased. Now it was just a matter of letting the plane lift off the runway. You did not have to pull back on the stick until you were airborne and wanted to set the climb attitude.

On to Norfolk

After takeoff, it was wheels up, reduce throttle and prop speed to climb settings, and set your course. Harry took off first, and I quickly rendezvoused with him. He picked up the radio signal from Jacksonville and set our course to follow the beam to the next checkpoint along the way. In those days we simply followed the low-frequency range station beams (no longer in existence) from station to station along the route.

As I recall, we followed the coast northward: Savannah, Georgia: Charleston, South Carolina: Wilmington, Delaware; then direct to Norfolk. I was flying loose formation on Harry, allowing me to look around a bit. But, as we approached Norfolk, he had me pull close in so we would look good around the Naval Air Station.

Harry was doing all the radio communication, got landing instructions, spotted the runway, and we made an upwind pass over the runway with me in right echelon as Harry made the standard break to the left. I waited five seconds and then made my break to establish a landing interval.

We were happily motoring downwind, wheels down, flaps down, slowing down to landing approach altitude and speed when the tower informed Harry, "Corsairs making landing approach. This is Norfolk City Airport. You are not cleared to land here. You were cleared for the Naval Air Station, across the river."

Yoiks! Harry had picked the wrong airport! He pulled up his wheels and flaps, added power and headed across the river. I followed a distance behind him as we flew just a few miles away to the now-clearly-visable Norfolk Naval Air Station runways.

That's Not All!

But there's more. Having had a problem in flight training one time getting my wheels down in a Corsair (I had to use the compressed CO2 bottle to blow them down), I decided to leave my wheels down. Harry had retracted his.

Now Harry contacted the tower at the Air Station, and got permission for a long straight-in approach, with me to follow. As Harry approached the runway, I was above and behind him, but the sun was in a position that showed me the shadow of Harry's plane—and he did not have his wheels down! I waited as long as I could to be sure he would let them down. I also expected to hear the tower tell him his wheels were up. They didn't, so I finally did. I shouted on the radio, "Harry, your wheels are up! Wheels are up!"

Harry took a waveoff while I came in and landed—and my tail wheel stripped! The solid-rubber tire simply shredded and we were stuck an extra day in Norfolk waiting for a replacement wheel. Harry was not a happy camper!

It's now 47 years later, and Harry doesn't even remember who I am. I called him a couple of years ago at his home in Texas for the first time in over 45 years. From our one-sided conversation, it was clear he had no memory of me, or of the incident I've just described. . .

CHAPTER 12

"Paddle Paradox"

by Fred "Crash" Blechman

This may seem absurd or unbelievable, but it's a true story—although I know of no documentation to back it up, and the details are from my recollection of a peculiar event over 47 years ago. . .

Fighting Fourteen (VF-14) with our F4U-5 Corsairs was on a short LANTFLEX (Atlantic Fleet Exercise) cruise in the Caribbean. We were the only full squadron aboard the U.S.S Kula Gulf (CVE-108), a jeep carrier that was one of several carriers participating in this annual wargame exercise. One of the other carriers was a British carrier with several squadrons aboard, including a Corsair squadron.

The Kula Gulf was cruising along that afternoon with a ready deck, flight operations on standby, trying to keep ahead of an approaching storm. I was one of the standby pilots in the ready room when the Duty Officer got a phone call that a flight of eight errant Corsairs—not from our carrier—were requesting permission to land. According to their radio transmissions, these were Royal Navy Corsairs whose ship was in the storm, and they needed to land somewhere, and they found us!

Since all our planes were parked forward on the straight-deck Kula Gulf, there was no reason we could not take them aboard. The LSO (Landing Signal Officer) hustled to his aft port station with his two landing signal flags ("paddles") while many of us went up to the "Vulture's Nest" observation platform on the ship's island near the bridge. We wanted to watch the Brits land those F4U-4 Corsairs, earlier lighter Corsairs than the -5 models we flew.

The British Corsairs, in two divisions of four each, made the standard upwind peel-off on the starboard side of the ship. Each established an interval as the pilot reduced rpm and manifold pressure, dropped the wheels and flaps, made a 180-degree turn to the downwind leg while losing some altitude, and then turned toward the ship while adjusting the nose attitude, airspeed, altitude, and bank to position the Corsair over the ramp with little or no straightaway.

This lack of straightaway was a necessary evil in a Corsair with a long tilted-up nose during the final approach. If you wanted to keep the LSO in sight, you had to watch him from the left side of the Corsair's nose.

The LSO's signals, except for the "cut" or "waveoff" were advisory, not commands. If you were OK, the paddles were held straight out for a "roger." If he held his paddles ABOVE the horizontal, he was advising that "you are high" and therefore should go lower. If the paddles were held BELOW the horizontal it meant "you are low" and should get higher.

There were also advisory paddle signals for approach speed and lineup with the center of the deck. Based on the LSO's paddle signals, you adjusted your flight path accordingly. You had to trust his signals since he was on the ship and could best judge the roll and pitch of the landing area as the carrier plowed through uneven seas.

However, as we watched the British Corsairs approach, something was wrong. Starting with the first one, they all were turning okay, but their altitude control was a paradox. When they were getting a "you are low" signal, they went lower, and when they were getting a "you are high" signal, they went higher! They all got waveoffs their first time around. Some got aboard on their second try, but most took three tries. Eventually they all got aboard. "Wow," we thought, "these guys must have had very little carrier landing practice."

We were wrong. Actually, these pilots were excellent carrier pilots—if they were following their own English "Batman" (their name for the LSO) signals. It turns out, from conversations with them after they landed (and after they stopped cussing "that bloody crazy Batman!") that British paddle signals for altitude were exactly the OPPOSITE of ours. To them the arms-up paddles signal meant "go higher" and arms-down paddles meant "go lower"! No wonder they were all over the sky until they figured it out. . .

CHAPTER 13

"Confessions of a Japanese Ace—How I Downed Five Corsairs!"

by Fred "Kamikaze" Blechman

Now that I have your attention from the title, "Confessions of a Japanese Ace—How I Downed Five Corsairs!", let me set a few facts straight before I get into the details. . .

For one thing, I was not in the Japanese Navy or Japanese Air Force during World War II. So I never actually received recognition for downing five Corsairs. I never got a medal, commendation, hero's welcome, or even a higher rank. I never was invited to dinner by Emperor Hirohito.

Maybe that's because I was a U.S.Navy fighter pilot, and World War II was over! But, IF I had been a Japanese pilot, and IF the war was still underway, and IF the Japanese heard about it, I would probably have been considered a Japanese ace. After all, by the time I got finished with those five Corsairs, they couldn't fly again until they were repaired! And downing five "enemy" aircraft makes you an "ace," right?

Why am I telling you this, over 40 years later? Well, if you are a pilot, perhaps you can learn from my mistakes. If you're not a pilot, maybe I can impart some vicarious thrills!

One Down!

It was early in 1950, after World War II, but just before the Korean fracas. I was in the final stages of flight training in the U.S.Navy, soon to get my "Wings of Gold."

I had gone through "selective flight training" in Dallas, Texas, soloing in a Stearman N2S "Yellow Peril" biplane after only eight dual hops, and hadn't hurt an airplane yet. The dark green grass stain on the bottom of the lower left wing from a near ground loop went unnoticed. So I was sent on to Basic Training at Pensacola, flying North American SNJs. For over flying 200 hours from solo through formation, cross-country navigation, gunnery, night flying, air combat, and even six arrested carrier landings, I scratched neither myself nor an airplane.

Then to Corpus Christi for Advanced Flight Training, flying F4U-4 Corsairs. In a Corsair, the first flight is solo. Then formation, cross-country, gunnery, dive bombing, night flying, and some air-to-air combat. I had no accidents, though accidents were common in this airplane, commonly called "The Ensign Eliminator." I was still a NavCad, with almost 20 months in flight training, now approaching the "final exam" of six arrested carrier landings in a Corsair. For this we were sent back to Pensacola.

As you can imagine, landing on a moving carrier deck requires special techniques. Flying tail-draggers, we had been trained from early on to make nice, stalled, 3-point landings. But the approach to a runway is quite different from the approach to a moving carrier. So you practice (and practice, and practice) by doing FCLP—Field Carrier Landing Practice. Using a marked field and an LSO (Landing Signal Officer) waving his "paddles," we would come around low and slow, follow the LSO's paddle signals, and get a cut or waveoff. While the field wasn't moving, and the landing was touch-and-go rather than arrested, this was reasonable practice.

Of course, we had done FCLP in SNJs before those six carrier landings that "graduated" us to Advanced Training. But the SNJ was relatively slow, with a short nose, good visibility, and only a few hundred horsepower swinging a two-bladed propeller. Low and slow was tricky, but controllable.

The F4U-4 Corsair was another story. The F4U-4 used a 2000+ horsepower engine swinging a giant four-bladed prop, and a nose extending 15 feet in front of the pilot. Flying low and slow, necessary in a carrier approach with the old straight-deck carriers of the day, your nose was up 10 or 15 degrees, completely blocking your forward view through the windshield. The approach had to be in a constant left turn, looking out to the left of the windshield. There was practically no straightaway before the cut.

I don't recall if it was my first pass, but I know it was the first day of Corsair FCLP. I came around, fighting the sluggish controls, and the large amount of right rudder necessary at that low speed. To add to the fun, the ground we were flying over at this outlying FCLP field was marshy, making the air very bumpy with rising thermals on hot days—and it was a hot day. My approach was almost good enough for a cut, but as I got close to the LSO I was getting too low and began skidding to the right. I raised my nose a tad and lowered my left wing to correct, but got a waveoff at the last moment. Feeling I was too low and slow, I added power quickly. Too quickly. Ever hear of "torque roll?"

In most American aircraft, the propeller turns clockwise if you're looking from the cockpit. As the blades push against the air, the air pushes back, trying to twist the prop counter-clockwise. This twisting motion is transferred back through the prop shaft, then through the engine, then to the fuselage and entire airframe. The plane tries to roll to the left. If you are already low and slow, with the left wing down, and suddenly add a slug of power, right aileron and right rudder take time to counteract the torque roll. I was too low, and my left wing hit the runway!

I yanked off all power and got my left wheel on the deck, instantly followed by my right wheel, and rolled to a stop. About three feet of the outer left wing was bent upward. Other than my pride, I was not hurt—but I never torque-rolled again! Lesson #1: Don't add power too quickly if you're low and slow.

Two Down!

After a Review Board Hearing, the Navy decided to let me continue toward getting my wings. Accidents happen. So I went back to FCLP, making something over 100 passes before the big day—six arrested Corsair landings aboard the U.S.S. Wright (CVL-49), a light carrier used by the Training Command at that time. A small group of us took off from Pensacola early that morning to be over the Wright, leisurely cruising in the Gulf of Mexico, at 9AM. As our flight approached, the Wright turned into the wind with a ready deck.

We formed a right echelon, flew upwind along the starboard side of the ship, and broke to the left in intervals and headed downwind to make our landings. One by one we made our passes, and got cuts or waveoffs. When a plane made a landing, the hook caught one of the eight (as I recall) arresting wires and brought the plane to an abrupt stop. The barriers (cross-deck cables held up by stanchions) forward of the plane were quickly dropped, and the plane deck launched from where it was. Then the barriers went back up for the next landing.

I made four passes, and got a cut on each. Only two landings more to go and I'd have those coveted gold wings I'd been struggling to earn for 21 months. My fifth pass was normal and I got a cut. I pulled off all power, lowered the nose, pulled back into a three-point position, and hit the deck. I could feel the hook catch a wire as I was thrown forward against the shoulder harness—but the plane's right wing suddenly dropped part way to the deck. The crash siren sounded as deck hands came running to the plane. My right wheel strut had broken!

The "crash" counted as an arrested landing. Of course, there was another Review Board Hearing. Was this pilot error, or had the wheel been weakened by so many Training Command carrier landings? This plane was regularly used (and abused) for FCLP and carrier qualification. I don't recall the final determination (probably pilot error), except that about two weeks later I got the go-ahead, made my sixth landing, and received my Naval Aviator wings and Ensign bars. I never broke another wheel. Lesson #2: Even a good landing can end up as an accident—sometimes for uncertain reasons.

Three Down!

I was assigned to Fighter Squadron Fourteen (VF-14), based at Cecil Field near Jacksonville, Florida, for my tour of duty in the fleet. As the Junior Ensign, I was always the last to take off and land. I also had to get familiar with the F4U-5 Corsair, a much-changed advanced model of the Corsair with many new features—and weighing an additional 500 pounds. Some "engineering improvements" added to this model of the Corsair were unnecessary, others were dangerous. But that's another story. . .

There was still the long nose, which required "S-turning" while taxiing to be able to see ahead of you. You would turn to the right so that you could see ahead by looking out to the left of the windshield, or turn left and look out the right. When making a runway landing approach, normally your nose was down in your descent, and you could see straight ahead, but as you pulled up your nose to flare out for a three-point landing, forward vision disappeared. As soon as the plane was rolling down the runway and under control you could begin S-turning.

One night we were practicing night landings at Cecil Field. As usual, I was the last plane to land. As I was making my approach to line up with the runway lights, I saw the Corsair ahead of me land and go rolling down the runway. As I pulled my nose up to flare out, I lost sight of the plane ahead. I was "hot"—coming in a little fast—so I touched down long on the runway, and immediately was concerned that I might be overrunning the Corsair ahead. As I looked to the left of my nose, watching for him to appear on the turn-off taxiway, I wondered if he had ground-looped ahead of me, and if I was about to crash into him. So, even though I was still rolling out pretty fast, I began applying alternate wheel brakes to S-turn so I could see ahead. First right brake to swing the nose to the right, then quickly left brake to keep from running off the right side of the runway, then right brake again.

Somewhere in that sequence I must have hit both brakes at once, or perhaps hit opposite brake while the plane was still turning, causing forward motion to suddenly decrease. For all intents and purposes, with the enormous inertia of the heavy plane, the center of gravity of the aircraft was now rotating around the wheels. My Corsair nosed down and the tail rose about 30 degrees! All four blades of the prop bent back, the engine ground to a stop, and the tail slammed back to the ground! Very embarrassing, since this was purely pilot error.

I never nosed-over again. Lesson #3: Don't S-turn at high speed!

Four Down!

During the two years I spent in VF-14, we went on about 12 carrier cruises. Two were six-month tours in the Mediterranean with the Sixth Fleet. But several cruises were Atlantic Fleet Exercises (LANTFLEX) usually held in Caribbean waters.

It was on one of those short cruises that we were flying off a small CVE "jeep" carrier, commonly called "escort" carriers. Considerably smaller than CVL or CV carriers, the jeeps had less landing area, less wires—but just as many barriers. On one flight I came back aboard normally enough, caught a wire, and watched the barriers in front of me drop.

Following normal procedure, I allowed the arresting wire to pull me back so a deckhand could release the hook. Now it was up to me to quickly get ahead of the barrier locations so the plane behind me would have a "clear deck." It was a matter of pride to make a snappy taxi forward to the parking area.

I raised the hook lever, hit the wing-folding control, and pushed the throttle forward briskly to get moving. As I rolled over the barriers, I cut the throttle and started applying brakes so as not to plow into the aircraft parked just ahead. The plane did not slow down. It appeared I had no brakes! Putting more and more pressure on the brake pedals until I was practically standing on them, and with the throttle completely off, I continued moving forward, as if I were on ice! The plane slowed down as I skidded forward, but didn't stop until the still-turning prop chewed off the sheet-metal tailcone of the plane parked ahead of me!

The propeller was not damaged. The Corsair ahead of me needed a new tail cone fairing before it flew again. Other than another blow to my pride, I was not hurt, but puzzled.

What caused the accident? Bad brakes? No! The brakes were fine. Too much throttle for too long? In a way, yes! You see, while our division had been out on patrol, the ship had gone through a rainstorm and the deck was wet. The rain water, mixed with the typical oil and gas on the deck, made the deck very slippery. I should have been informed of this. I wasn't.

Still "pilot error." I didn't do that again. Lesson #4: Before adding power to taxi, make sure you can stop before hitting any obstructions ahead!

Five Down!

Some months later, on another LANTFLEX, and flying from another CVE, I had an early morning search flight. Hours later I was called on deck to taxi my Corsair out of the way for other planes making an unscheduled launch. But one of the planes intending to fly had engine trouble. I was the closest replacement airplane and pilot, so I was put on a catapult and shot off as part of the flight.

We flew around for several hours looking for an "enemy" patrol plane reported by radar, but never found it. The deck was spotted for launching a scheduled flight, so we had to tool around, tail-chasing, until the deck was clear for us to be recovered.

Hot and tired after two long flights, I made a normal approach and was so relieved to get the cut that I relaxed and "dove for the deck" (let the nose drop too far before pulling back.) My Corsair main gear hit the deck and bounced the plane back into the air, flying over several arresting wires in the process. I instantly realized this, dropped my nose, then quickly pulled back, and caught the last arresting wire. Unfortunately, on this small ship the last arresting wire allowed the heavy Corsair to reach the first barrier cables, which bent two blades of my prop.

I never did that again, either. Lesson #5: Don't relax until you are in the chocks with your engine off and the prop stopped.

Epilog

I wasn't hurt in any of these accidents—not a scratched finger—and never had another flying accident, although I got a Commercial Pilot's License after getting out of the Navy. Four of the five accidents were carrier related—an especially accident-prone environment in the days of the straight-deck carriers—and all were in Corsairs.

I learned something from each accident. Not a nice way to learn, but effective—if you survive. . .

ACCIDENT REPORT #3

Date: 21 November 1950, 22:12

Pilot: ENS Frederick Blechman USNR

Organization: VF-14, NAS Jacksonville, CVG-1, ComFairJax, CNO
Aircraft: F4U-5 #121905
Purpose: Night Familiarization
Hrs.last 3 months: 59.3; Total hours: 362.7
Location: NAS Jacksonville, Runway 31
Weather: VFR
Injuries: None

ACCIDENT:
Pilot had been airborne one hour and 27 minutes. The first hour was spent on airwork familiarization and then the four aircraft on this flight returned to base to practice touch and go landings. Blechman made five touch and go landings then came around for his final landing. He made an approach and landing, and during rollout Blechman unlocked his tailwheel to swing plane slightly to the right to observe the runway ahead. During this maneuver right swerve developed. In attempting to stop swerve, pilot applied excessive left brake causing plane to nose up. Prop tips struck runway, then plane came to stop and settled back to 3-point attitude.

SPECIFIC ERRORS:

(1) Attempting to control aircraft at relatively fast speed during landing rollout with tail wheel unlocked.

(2) Applying brakes simultaneously with excessive pressure, causing plane to nose up.

ANALYSIS:

Runway Duty Officer, whose responsibility it is to warn pilots of wheels-up approaches, simultaneous approaches at too close intervals, plane dead on runway, etc., observed Blechman's approaches and landings. He observed that Blechman's landing touchdowns appeared to be slightly longer on runway than other planes. This situation was probably not noticeable to pilot on his "touch and go" landings, as he was immediately applying throttle after "touchdown,' and was quickly airborne again. While making his final landing, Blechman was of the opinion that another aircraft was still on the runway and started the maneuver described which developed into the accident. This squadron has instituted the practice of having senior officer on downwind end of runway when new pilots are conducting night familiarization landings to observe and coach if necessary, to prevent recurrence of such accidents. Another practice that has been adopted is that each landing aircraft call the tower and report when clear of the duty runway. It is recommended that other squadrons consider this policy.

COMMANDING OFFICER:

No disciplinary action was taken since it was considered that the pilot erred in judgment rather than through negligence.

REMARKS:

Damage:Prop blades.

ACCIDENT REPORT #4

Date: 7 November 1951, 15:10

Pilot: ENS Frederick Blechman USNR

Organization: VF-14 USS Kula Gulf, CAG-1, ComFairJax, ComAirLant, CNO
Aircraft: F4U-5 #122158
Purpose: Combat air patrol
Hrs.last 3 months: 101.2; Total hours: 666.2
Location: USS KULA GULF (CVE-108)
Weather: VFR
Injuries: None

ACCIDENT:
ENS Blechman was the number four man in a flight of four F4U-5s. Upon completion of a routine combat air patrol flight the division of aircraft entered the pattern for a normal breakup and carrier landing. ENS Blechman made a normal approach and was in a good position when given the cut. After taking the cut, ENS Blechman nosed the aircraft over and dived for the deck. The aircraft hit wheels first and bounced back into the air. The hook hit the deck after the number one cross-deck pendant and then the tail bounced back into the air. The aircraft traveled up the deck engaging number eight cross-deck pendant and hit number 2 and 3 barriers.

ANALYSIS:

It is the opinion of this board that ENS Blechman erred by the fact that instead of making a normal landing, he dived for the deck and bounced back into the air causing him to catch a late wire and engage the barriers.

CONCLUSION:

The primary factor causing this accident was the pilot's error in diving for the deck. This board recommends that all flight squadron commanders remain cognizant of the points brought out in this aircraft accident report. The technique involved during this critical period between the "cut' and the arrested landing should be periodically reviewed by all squadrons for the respective type aircraft.

COMMANDING OFFICER:

Concurs. All pilots are briefed immediately following each carrier recovery regarding individual pilot technique. Those pilots not flying observe launches and recoveries.

Following this accident a movie camera was set up to take pictures of each pass, starting at about the time the pilot is picked up by the LSO. These movies are reviewed each night by all pilots, with emphasis placed upon the landing. This procedure is recommended for squadrons to use when field qualifying. Also, when field qualifying, the squadron commander and the LSO must place a great deal of emphasis upon the pilot technique in landing the aircraft following the cut.

REMARKS:

Damage: Prop was damaged and a few dents and tears inflicted to the speed ring, necessitating a change of both parts.

CHAPTER 14

"Carrier Crash!"

by Fred "Crash" Blechman

It has been said that the most dangerous time in a pilot's career is when he has about 600 flying hours. Prior to that time he's very careful and deliberate. After about 600 hours flight time he tends to be more relaxed—and gets careless. I had 666.2 hours of flight time, with 454.6 hours in F4U Corsairs, when I crashed on the deck of an escort carrier!

It was a bright, clear dawn in the Caribbean on November 7, 1951 when eight of us in Fighter Squadron Fourteen (VF-14) were shot (they called it "catapulted"!) from the escort carrier U.S.S.Kula Gulf (CVE-108.) Our F4U-5 Corsairs were part of an annual training exercise called "LANTFLEX" (AtLANTic FLeet EXecise.) We were the Red Squadron, flying CAP (Combat Air Patrol) to protect our small task force from any Blue (enemy) raids.

Nothing special happened. We just flew around in large, lazy circles in loose formation over the endless sparkling water, some distance from the carrier and its support vessels. I was flying F4U-5 Navy Serial #122158, Squadron #405.

After over two hours of occasional vectoring by the carrier CIC (Combat Information Center), we headed back to "home," flying in right echelon past the starboard side of the carrier's island as we peeled off to port, setting our landing interval. We landed in turn without incident, and headed for the ready room. The Acey-Ducey ("Backgammon" to landlubbers) and card games came out, and we relaxed. I was not scheduled for any other flights that day after our early launch and relatively long 2.6-hour flight.

It was late morning when things changed suddenly! Our radar had spotted a "snooper," apparently a Blue patrol plane approaching our ships. "Pilots, man your planes!" was called for those scheduled on standby. Although I was not scheduled to fly, our deck was not spotted for the unexpected launch, so I went up on deck in case I was needed to taxi a plane to a new position on deck.

It soon became apparent that some of our planes would have to be moved. I climbed into the same #405 I had flown earlier, just expecting to taxi around the flight deck as directed during respotting aircraft. I had my regular flight gear—a hardhat, G-suit and parachute, standard procedure in case of a standby launch—but no plotting board, and no briefing.

This was to be a four-plane search-and-destroy mission. Three got off fine, but the fourth had engine trouble. All planes were being catapulted since the wind over the short deck was not sufficient for a safe deck launch. (Not that cat shots were all that safe!) They took the sputtering dud Corsair off the port catapult, put me on, hooked up the shuttle and cable, and shot me into the gathering clouds! Equipped with an extra gas tank, we were off for a three-hour search flight.

This turned out to be a long, boring, very tiring flight. The flight leader, to make things more interesting, put us in a "tail chase"—and I was the last plane in this whipping tail as the leader performed mild aerobatics. The idea was to stay in position behind the plane ahead of you. Following was relatively easy if you were in one of the up-front positions in this tail chase, but got progressively more difficult if you were further back in the stack. I was in position #4, the end of the tail, and was using lots of throttle, rudder, elevator and aileron movement, trying to stay in position. (This wasn't as bad as being in the #8 position in a tail-chase, as I had been a number of times, but it was grueling nevertheless.)

The F4U-5, the heaviest in the Corsair series, did not have boosted controls, and didn't need them for normal flight. But it took a lot of physical effort to horse it around the sky. Also, we had gone up above the cloud layer, and the sun was beating through the bubble canopy. Combined with the natural high humidity of the Caribbean, the inside of that bird was h-o-t and s-t-i-c-k-y! I recall popping the canopy back a few inches several times to try to cool off.

Finally, after three hours, we were called back to land. There had been another unscheduled launch while we were airborne, so now the deck had been respotted again for our recovery. These were still the days of straight-deck carriers, when reshuffling of planes on deck was a common and necessary procedure between launches and recoveries.

We spotted the Kula Gulf, steaming ahead of its bubbling, churning wake, surrounded by several smaller support vessels and their smaller, shorter white tails contrasting against the shimmering sea. A rescue helicopter, always aloft during air operations, hovered nearby.

As we approached the landing pattern in right echelon formation, flying upwind along the starboard side of the carrier for the break-off, I reflected about how well I had been doing. I mentally patted myself on the back for my good ordnance scores, and, although there had been a rash of accidents on this cruise, my slate was clean.

Landing an F4U-5 on a small escort carrier was inherently marginal. Escort carriers (CVEs), with a flight deck under 600 feet long, were small compared to the larger 800- and 1000-foot light (CVL) and battle (CV) carrier decks. Escort carriers had fewer arresting wires (eight, compared to eleven for CVLs and thirteen for CVs, as I recall), and their top-heavy decks on small hulls had a much greater tendency to pitch, yaw and roll even in light seas. Every landing was a challenge.

As I peeled off to the left and set my interval for the downwind leg, I looked forward to getting down. I was very tired and sweaty. Getting back on deck, into a shower, and then sacking out—that's what I was planning.

I dropped my wheels, flaps and hook on the downwind leg, throttled back to lose some altitude, and used the ship and its wake to judge my abeam position, direction and altitude. The ship was steaming upwind, and I was flying downwind, so it took no time at all before it was time to turn left onto the base leg.

I pulled back on the throttle, slowly dropping altitude on the base leg by referring to where the horizon cut the bridge, finally settling at the approach altitude and maintaining just enough power to hold the nose-up attitude at about 90 knots, hanging on the prop. I put the left nose of the Corsair on the aft starboard deck for an intercept course and held it there. As the ship moved forward at about 20 knots, I pulled the Corsair around to the left, watching the LSO (landing signal officer) for paddle instructions.

There was no luxury of any significant straightaway in landing on those old straight-deck carriers when you were flying a long-nose Corsair in a nose-up attitude. You just couldn't see ahead of you—only off to the side. We essentially pyloned counterclockwise around the LSO in order to keep him in sight at his port fantail location.

As I got close in, I pulled the nose left toward the ship's centerline. This was effected by the wind over the deck, which was never straight down the deck, but about 15-degrees to port so the turbulence from the ship's stacks and bridge did not appear in the flight path of the landing planes. This made for a very tricky last few seconds. . .

At this slow speed, just a few knots above stalling, it took a lot of right rudder, even though in a left turn. And you didn't dare add power quickly—even if you thought you had to—since the 2300 horsepower engine turning the 13-foot diameter four-bladed prop would make the aircraft roll uncontrollably to the left—the dreaded torque roll.

It took a lot of back stick, considerable power, and almost all my right rudder to hang in there. As I approached the ramp in a left turn, the LSO's paddles and my own perception was that I was drifting to the right of the deck centerline. Too much right rudder. I cross-controlled a bit and slipped to the left just as I approached the ramp, leveled my wings, and got a mandatory "cut."

"Ah, home at last," I thought as I relaxed, dropped the nose, and pulled back to drop the tail so my hook would catch an early wire. But I relaxed too soon! Perhaps I was more tired than I realized, and didn't pull back soon enough, or perhaps the deck lurched up at that time. Whatever the reason, my wheels hit the deck and bounced. I was flying over the arresting wires, tail up, and drifting to the left! (Photo #1.)

I heard the crash horn just as I popped the stick forward (Photo #2) to get back on deck, and then quickly pulled back (Photo #3), to get my arresting hook down. I caught #8 wire—but on this ship with a heavy Corsair the arresting cable pulled out just enough for the prop to catch the uplifted barrier cables. Strike two prop blades! (Photo #4.)

The moral of the story? Don't relax at the wrong time. The flight isn't over until the wheels are in the chocks and the prop (turbine?) has stopped.

November 7, 1951: Author's F4U-5 bounces and begins to fly up the deck of the jeep carrier, USS Kula Gulf (CVE-108) . . .

Wheels are back on the deck. Now it's time to get the tail down so the hook can snag a wire . . .

The hook catches the last wire, #8 on this carrier—but the barrier cables loom ahead . . .

OOPS! The arresting gear wire stretched out enough for two blades of the prop to engage the barrier cables.

CHAPTER 15

"WAR DOG—The Ten-Engine SNJ"

by Fred "Crash" Blechman

It was just a hunch. I was watching a yellow-nosed SNJ doing an aerobatic performance at the Santa Barbara Airshow in 1993. Twisting and turning, yanking and banking, with popular West Coast airshow performer John Collver at the controls, this SNJ was pirouetting around the sky with the engine alternately growling, screaming, or completely silent! This plane was doing things I didn't recall ever learning when I was in Navy flight training flying SNJs over 40 years earlier. "Hmmmm," I wondered, "is it possible that this is an SNJ I flew back in those days?"

So, when John finished his whirling-dervish performance, got his accolades from the crowd and taxied to his parking chocks, I sauntered over to this two-place tandem trainer with its multi-ribbed double sliding canopy.

This silver SNJ-5 was in mint condition, with a custom huge, shiny, silver dome-shaped propeller spinner in front of the yellow cowling. It was painted in the colors of VMT-2, a Marine training squadron that flew out of the Marine Corps Air Station in El Toro, California, during World War II. The markings were black with green trim and yellow wheel hubs, with the number 17 and a big WD on the vertical fin and rudder. Right behind the cowling were the words "WAR DOG." The airframe serial number, 90917, was clearly marked under the empennage. I wrote the serial number down to check later.

I approached John Collver, the pilot-owner of WAR DOG, who looks the part of one of "those magnificent men and their flying machines," sporting a large flowing mustache with a big smile, and wearing a green flight suit. I asked him about the history of this particular airframe, wondering if it had ever been a Navy trainer at Pensacola.

He told me he had all the airframe logs, and that WAR DOG had, indeed, at one time been stationed at Pensacola. He also pointed out that it was (at the Santa Barbara airshow) on its NINTH engine, having worn out the eight previous engines!

Hmmmmm. So it HAD at one point been based in Pensacola as part of its colored past. After the airshow, I hunted down my Navy Pilot's Log Book, which included flight training in SNJs and F4U-4 Corsairs, as well as my fleet squadron time flying F4U-5 Corsairs with the VF-14 "Tophatters." Looking down the log of training flights while in Pensacola—sometimes several in one day—I checked each serial number of the aircraft flown. Hmmmm. Lots of serial numbers around 90917, but no 90917 so far. . .

There it is! On February 5, 1950, I flew this exact airplane from Pensacola to New Orleans on a cross-country solo flight! This was not just deja vu—I flew this plane over 49 years ago, before the present pilot-owner was born!

Well, of course, now my goal was to fly in this airplane again. It took over a year until John Collver was performing at an airshow I attended—Torrance Air Fair '94. He told me the plane recently had another new engine—the TENTH on this airframe—because of some contaminated gas at the Watsonville Fly-In earlier in the year. That gas wrecked his Engine #9—and many others!

I made arrangements with John on Air Show Preview Day, and he graciously took me up in the back seat for a 25-minute flight around Long Beach, Palos Verdes and the Southern Los Angeles area—with a slow-roll thrown in—while I snapped photos and swiveled around in my shoulder straps trying to take in everything with my camcorder. There's the Queen Mary and the dome where the Spruce Goose used to be! Look at Long Beach, and downtown Los Angeles, and Century City's towers in the background! There's the Goodyear Blimp flying nearby—but John would only get within about 1000-feet of it.

It was like I'd never flown around Los Angeles before. Somehow it was different looking through the bird-cage canopy cruising at about 145 knots and 1500 feet altitude, compared to peering out a small airliner window while zipping over the landscape at hundreds of miles an hour.

I'm 71 years old now, and have flown many different airplanes, from prop-driven fighters to ultralights to canards, but I guess I'll never outgrow my fascination with watching the ground scroll by from an airplane, zipping right above a cottonfield of clouds, dodging between cumulus cloud puffs—or making the whole world tilt by just pushing the stick to one side. . .

Sidebar 1

Well, I thought that coincidence of finding and flying in WAR DOG after 44 years was great—but this turned out to be just the beginning of a whole series of "throwbacks!"

While at the November 1994 AvCad/NavCad Reunion in Pensacola and touring the National Museum of Naval Aviation, I photographed an N2S Stearman "Yellow Peril" with a side number of "41," identical to the side number "42" I flew on my first solo flight. Later on the museum tour, we found ourselves on a replica of the deck and island of the U.S.S. Cabot (CVL-28), the carrier on which I made six arrested landings in an SNJ-5C on March 23, 1950 while in basic flight training to qualify for advanced training. At the Reunion Airshow, one of the "performers" was an F4U-4 Corsair, the type I flew in advanced flight training, and in which I had to make six arrested carrier landings to earn my "wings of gold."

On the way back to California from Pensacola, we stopped in New Orleans and took a short river cruise. I couldn't believe it! Moored on the Mississippi River, being prepared as a museum attraction, was the actual U.S.S. Cabot on which I had made those six qualifying SNJ landings in 1950!

When I got home in the waiting mail was a copy of AVIATION History Magazine. On Page 4 there was a picture of "WAR DOG," #90917, which began this whole series of flashbacks. Weird, eh?

But that didn't end the series of coincidences. Around Christmas of '94, through a real odd set of circumstances, I found myself talking to a fellow I knew in 1946, but with whom I'd had no contact since. It turns out that he owns an SNJ based at Van Nuys Airport—8 miles from me—as one of the well-known "Condor Squadron" of AT-6s and SNJs. The serial number of his plane is 90918—which my logbook shows I flew on a solo air-to-air gunnery hop on 25 February 1950!

Very soon thereafter I received a catalog from a small mail-order company that was selling off some miscellaneous airplane-related items. One of the items was a signed limited edition painting of the U.S.S. Cabot, mentioned earlier. Of course, I bought it. When it came, a flyer was enclosed offering various military patches—including a rare U.S.S. CVL-28 Cabot patch. Naturally, I bought that, too.

As Yogi Berra, the famous former baseball player supposedly said, "It's deja vu all over again!"

Sidebar 2

The Pilot and the Plane

John Collver, 46, is now the pilot-owner of WAR DOG, together with his wife, Donna, their three small children, and Donna's parents, Ron and Joanie Custer. It is normally based at Zamperini Field in Torrance, California. John has been flying this plane for 21 years.

Like many airshow pilots, John has a "regular job" as a corporate pilot for the Northrop Corporation, with over 13,000 hours of logged flight time in over 50 types of aircraft, including the Goodyear Blimp. He began flying in 1968 at age 13 by washing airplanes in exchange for flight time and had soloed five different airplanes by the time he was 16—before he learned to drive a car. After instructing and flying char-

ters, he entered aerobatic competition, and performed in a Super Decathalon and a Great Lakes biplane before WAR DOG.

Oddly enough, WAR DOG was not always an SNJ-5. It first flew on November 6, 1944 as an AT-6D, built at North American Aircraft's Dallas, Texas, plant. It was later transferred to the Navy and was stationed at bases such as El Toro, Miramar, Pensacola, and others until it was retired in 1956. While facing the scrap pile at Arizona's Litchfield Park, it was rescued in 1952 as part of the Japanese Defense Force until 1972 when it was returned to civilian registry in the U.S.A. Later it was sold in scrap condition for $500 and restored at Warbirds West in Compton, California.

The airframe logbooks indicate WAR DOG has flown over 10,000 hours. The engine, a 600 horsepower Pratt & Whitney R-1340 Wasp, is the tenth engine this airframe has used. WAR DOG weighs about 5800 pounds, cruises at 170 mph, with a top speed of 240 mph and a full-flap stall speed of 57 mph. It uses fuel at the rate of 80 gallons an hour on takeoff, but only 35 gallons per hour in cruise. Collver estimates the total cost of operation is $200 an hour.

For further information on John Collver's performances, call (310)539-3640, or write Warbird Air Shows, 2456 West 247th St., Lomita, CA 90717.

CHAPTER 16

"Finding Your Lost Flying Buddies"

by Fred "Crash" Blechman

Donald J. Flynn, where are you? Remember me? We were carrier pilots back in the early 1950s, flying F4U-5 Corsairs with Navy Fighter Squadron Fourteen (VF-14). I recall times we went out on the carrier catwalks at night in the middle of some ocean, looking at the millions of stars swaying in a jet-black sky as the carrier pitched and rolled—and remember the one night we thought we saw a UFO? I've been looking for you the last few years, Don. I've found—and written—28 different Don, Don J., Donald, Donald J. and D.J.Flynns all over the USA—but I'm still looking for the Ensign Don Flynn from VF-14.

And "Doc" Mossburg? Where are you? I was your wingman in VF-14. I've checked with Mossburgs all over the country, but can't find you, Doc. You were a LTJG in 1951, so you must be about 69 years old—if you're still alive.

Those were the "no shows." That's the bad news. But the good news is that I HAVE found almost 20 of the "old flying buddies" that I've been looking for! If you'd like to find some of YOUR old friends, schoolmates, flying buddies, shipmates, or military associates, I'm going to describe the methods I've used.

Making and Using a List

Before you can start to use your computer—or any other method—to find someone, you must first make a "target" list with as much information as you can find. Certainly, right off the top of your head you'll remember some guys (or gals) you'd like to locate. Start with those, then add to your list—and perhaps even find the last-known addresses and phone numbers—from class yearbooks, reunion lists, Navy cruise books, old address books—and referrals from those you do find.

If you are computer literate, and have the equipment , once you've made the list, you can use the Internet or computer CD-ROM discs to "home in" on likely prospects. It will take some persistence, phone calls, and post cards to "pin down" the latest location of these people. Then you'll call or write them and perhaps get together. Some will be very responsive, some not. Some will remember you clearly, some won't remember you at all!

I'm no professional tracer of lost persons, so I'm sure there are additional methods, but these are the ways I used my computer to find some "lost" flying buddies.

Class Books

Certainly, your high school and college yearbooks are a source for names and faces, since most have photos and home addresses. Even if you haven't personally saved them, you'll find some classmates that have—and that leads to reunions, and reunion rosters.

Some individuals take it upon themselves to become historians. When I was in the V-5 Naval Aviation College Program at Swarthmore College in Pennsylvania in 1946, the school yearbook, "HALCYON," included the names and addresses of all the V-5ers (and V-12ers) in attendance. Many of these guys went on with me to Dallas for Selective Flight Training in late 1946.

Bob Voiland, one of the V-5ers, sent letters to all on the roster and has compiled a 4-page single-spaced list showing locations and phone numbers for about 70% of them—an amazing feat when you consider that this is a list from 1946! Through this list I found Ara Martin Boyajian, former Swarthmore V-5er who made it through flight training, flew F8F Bearcats in the fleet, and now lives only an hour away from me!

Cruise Books

It's common in the Navy for "cruise books" to be produced on extended cruises. I was on two Mediterranean carrier cruises with VF-14 (USS Wright in 1951 and USS Wasp in 1952) and each had a cruise book. Typically, aside from the many pictures and stories about ports visited, there will also be a roster of personnel.

I still have those cruise books. They are a great resource for recalling old buddies, and provide a starting point for finding them. Of course, some have died, and most have moved. But many, after their military service, return to their home town, or nearby. Even if they don't, there are frequently relatives who are still in their home town that can lead you to the one you're seeking.

Some military units keep a history, and some individuals do it on their own. Former VF-14 "Tophatter" Robert Holmbeck has compiled a list of Senior Tophatters. Although this list included VF-14 pilots going back for many years before I joined the squadron, I located several of my old flying mates, among them "Cookie" Cleland, Jesse Hopkins, Don Ross, and Dale Fisher (Dale since deceased.)

When I called Don, whom I'd had no contact with in 42 years, I simply said, "Is this retired Navy Captain Don Ross, the world's best Corsair pilot?" His immediate reply, with no hesitation, was "Freddie Blechman!" He recognized my voice! Eerie!

Using Holmbeck's list and squadron history information, LT Paul McSweeney, at the time a rear-seater in F-14 Tomcats with VF-14 at Oceana, Virginia, coordinated a VF-14 75th Anniversary Squadron Reunion. I was there!. Wow!

Referrals

Once you find one "buddy," they may still be in touch with others. All you have to do is ask. That's how I found Carl and Merle Hilscher; I was in flight training with Carl, and used to date Merle before he married her!

I found Jim Gillcrist (pre-flight at Ottumwa, Iowa in 1946) through his brother, retired Rear Admiral Paul T. Gillcrist, who wrote a GREAT book, "Feet Wet" (perhaps the best flying book I've ever read!) Recognizing the last name, I wrote Paul and he put me in touch with his older brother, Jim.

I found my old VF-14 skipper, retired Captain Robert C.Coats (Navy WWII 10-plane ace) by contacting Tommy Blackburn (former WWII VF-17 Navy Captain, now deceased) after reading his exciting "The Jolly Rogers" book. Skipper Coats put me in touch with A.G.Wellons, who retired as a Navy Captain.

Computer Searching

I knew that my room mate throughout most of flight training, John "Jack" Eckstein, was from Massillon, Ohio. I kept meaning to try to find him (our last contact was over 20 years ago, when he was still a Commander in the Navy), but I kept putting it off. Finally, one evening about seven years ago, on a whim, I decided to give try to find Jack using my computer.

I dialed up CompuServe, typed my password, accessed PHONEFILE, and requested all ECKSTEINS in Ohio. 153 of them! Scrolling through, I found a John Eckstein in Massillon, Ohio—where Jack lived over 40 years ago. Worth a try. I called. The woman who answered turned out to be Jack's mother!

"Where does Jack live now," I asked. "Oh, he lives in Virginia. Has a vineyard," was the reply. "Can I have his number?" I asked. "Well," Jack's mother replied, "I can give you his number, but he's not there now. He's on a trip to Chicago. But it just so happens that he stopped by this evening to visit, and right now he's across the street visiting his sister—my daughter."

I got the number "across the street" and called. Can you imagine Jack's surprise? It was priceless! Jack retired from the Navy as a Captain and now raises grapes for local wineries. We've gotten together twice, with a third time coming up soon.

Another contact through CompuServe's PHONEFILE was Ed

Balocco. I went through flight training with Ed, who became a Marine fighter pilot, and I lost track of him. I knew he was from Antioch, California. Through PHONEFILE I found a sister, who directed me to Ed, now a lawyer in nearby Walnut Creek, not far from Antioch. When I called Ed and said, "Hi, Ed! This is Fred Blechman. Remember me?"

He hesitated for a moment then blurted out, "Freddie Blechman! Sure I remember you. I talk about you all the time!"

I was shocked. Talks about me all the time? Huh? I hadn't had contact with Ed in 42 years. Why would he talk about me? I asked him why. "Oh," he replied, "I tell people about the skinny kid from New York who went through flight training with me, then went on to fly Corsairs off carriers—and didn't know how to drive a car!"

It was true. Funny what people remember. I learned to drive a car after about 30 carrier landings. . . I flew up to Oakland to see Ed, and showed him pictures I took of him in flight gear 48 years ago. We had a fabulous personal reunion—and I got a fantastic personally-guided tour of the San Franciso area.

PHONEFILE found me another old friend. He and I used to go flying in rented planes when we went to school at Cal-Aero Technical Institute in Glendale, California in the 1940s. I found Ugo Sbaraglia in Santa Barbara. It was easy. He was the only SBARAGLIA in all of California! We've gotten together twice since—once at the annual Santa Barbara Airshow.

In more recent years, the Internet and CD-ROM "phone disks" provide access even more extensively. For example, without going into details, www.yahoo.com has a "people search" function that is fast and easy to use.

Homing Into Target—Follow-up

Once you think you've found someone on your list, you can call or write. Calling can be expensive, especially if you have a long list of "possibles." In that case, I send inquiry postcards, like "Are you the Don Flynn I flew with in VF-14 in 1950/51?" If I'm pretty sure I've got a "hit," I call and surprise them with a blast from the past!

Sometimes a phone call can be embarrassing. I found a former petty officer who was my right hand man when I was Education and Training Officer in VF-14. We worked closely together for two years. I called him. He didn't remember me. I wrote him. He never answered.

So be ready for some disappointments. Some you contact will not care at all. Others will be excited to get together and hash over old stories.

SIDEBAR

If you want to learn about more conventional ways—with and without a computer—to find someone, check out "FIND THEM FAST! A Guide to Finding Anyone Quickly, Cheaply and Easily," a self-help search book written by award-winning investigative reporter Dave Farrell.

During nearly 18 years working his investigative beat, Farrell picked up scores of search tips and techniques used by professional investigators, police officers and reporters. He has included many of them in this 114-page soft-cover book.

Twelve chapters discuss the details of tracing through credit reports, vehicles, voter registrations, corporate records, marriage and divorce records, land and tax records, and other means. Nine detailed appendixes list the actual addresses to contact for this information.

Use a nationally recognized non-profit organization to search the entire country for your subject for $10. Get the U.S.Postal Service to help you locate your subject for $5. Find anyone in any branch of the armed services. This book even tells you how to find people using hunting, fishing and other licenses.

Farrell also discusses the use of the Social Security Number for tracing an individual using an inexpensive computer data base—or to access the Master Death File with the names of 43 million deceased Americans, and using death records to find living persons!

If you are at all serious about finding someone, FIND THEM FAST! can be ordered for $9.95, postage included, from Dave Farrell, P.O.Box 252511, West Bloomfield, MI 48325.

PART II

"REMINISCENCES OF A NAVAL AVIATOR"

A GROUP OF SHORT STORIES

By Daniel L.Polino (c) 1987

(Published with permission)

Here is a group of short stories by former Navy fighter pilot, Dan Polino, as self-published in a limited edition 95-page book for his friends and family.

The book contains 56 stories in three sections: Cadet Days, Operational Flying, and Post War Reserves.

I've selected eleven of his many stories involving the F4U Corsair.

Dan is originally from Buffalo, New York, and currently resides with his wife, Cathy, in Simi Valley, California.

CHAPTER 17

"LOSING A SQUADRON MATE"

by Daniel L.Polino

When we arrived at the Wildwood, NJ. Naval Air Station, we began to train for the first time as a fighter-bomber squadron. At the time, we were using older versions of the F4U Corsair, some with the old "bird cage" canopies.

The airfield was surrounded by pine trees, and those nearer the runway were cut close to the ground. One phase of our training in Corsairs was night flying, and especially touch-and-go landings. This consists of approaching the runway, making a tail-first landing, and immediately giving it throttle and taking off again without rolling to a stop.

To accomplish this, half the squadron was given the evening off and went on liberty to the town of Wildwood, while the other half spent a couple of hours making touch-and-go landings. It was on the first night that I was in town with some of my squadron mates when word was received that one of our aircraft had crashed and burned.

When we returned, we found that, after one of our pilots had made his touch-and-go landing and applied power to take off again, he got as far as the end of the runway, climbed about 50 feet, retracted his landing gear, and had an engine failure! He had no choice but to belly-land straight ahead in the grass off the end of the runway.

His misfortune, however, was that his Corsair hit the stump of one of the trees and it gouged out his fuel tank. The resultant fire severely burned our squadron mate, especially where he was unprotected by his Mae West life jacket. He lived 24 hours, leaving a young widow.

The following night was our turn to practice touch-and-go landings. The landing gear lever on the older models was located on the side of the cockpit near your left foot (rudder pedal). It was a pivoted lever with a spring-loaded pin to set the gear in either up or down position. This required after clearing the runway for you to reach down by your left foot and pull the pin, then set the landing gear lever in the "up" position. Conversely, you'd lower the gear for landing.

However, my aircraft was so old and ill-repaired that, instead of a spring-loaded pin, it had a spike! You pulled it out so far, moved the lever, and shoved it back in to hold that position! It was bad enough having to duck down in the cockpit, loosening my seat belt to do so, during takeoff; but to wonder if the spike would hold its position was too much. I was glad to get that night over with and felt fortunate that I was able to survive. It didn't help much knowing that one of our team was, at the time, dying in the Navy hospital from the previous night's exercise. . .

CHAPTER 18

"HOW I NEARLY BROKE THE SOUND BARRIER IN AN F4U"

by Daniel L. Polino

While with fighter-bomber squadron VBF-152, under Commander DeVane, we had a lot of dive-bombing sessions. The F4U Corsair had been adapted for use as a dive bomber by modifying the main landing gear for use as dive brakes.

When the landing gear control was placed in the "DIVE" position, only the main gear dropped; the tail wheel and hook remained retracted. It worked fine. The main gear acted not only as a dive brake but, better still, provided automatic nose-down trim. Normally, as a plane builds up speed in a dive, you have to roll in nose-down trim for the elevators as the plane tries to climb at higher-than-cruise speeds. Conversely, when pulling out of a dive, simply raising the main landing gear retrims the plane to a nose-up condition; you don't even have to pull back on the stick.

When using the F4U Corsair as a dive bomber, due to its inherent clean trim and aerodynamic characteristics, the aircraft is flown during the dive with the main landing gear in the "DIVE" position. This limits your dive speed to approximately 350 knots, and allows you to maneuver during the dive in order to put your sight on the target.

During practice runs we'd climb to 12,000 feet and, from a right echelon formation, peel off at ten-second intervals and follow each other in a steep (70-degree) dive to a floating target just off the ocean coastline. In spite of the dive brakes, the Corsair does reach a high speed rather quickly and, as a consequence, the bombs must be dropped and a pullout initiated soon after 3000 feet of altitude.

On one flight that I'll not forget, I found myself flying as wingman on the squadron commander. This meant that I would be second man into the dive. We were carrying 1000 pound bombs at the time. As the commander peeled off for his dive, I noticed he was NOT extending his dive brakes! Following him ten seconds later, I elected to dive "clean", as he did, otherwise I might be overrun by the man behind me if HE chose to follow the commander's clean dive configuration—and I didn't. . .

Diving the F4U Corsair, even with the engine in the idle condition, without the landing gear acting as dive brakes, causes a rapid buildup of speed in steep dives of this type. As I approached the floating target just offshore, I was so intent on getting my aiming point set relative to the target that for a moment I didn't look at the air speed indicator. However, the noise of the wind whistling through my wings brought it to my attention. At 6000 feet, I was reading over 420 knots indicated and increasing. If you take into account the conversion to miles-per-hour, the lag in the instrument reading, and altitude pressure corrections, you approach 500 miles per hour air speed! I also realized that the moveable members of the wing and tail surfaces were fabric-covered.

At that point, I released the bomb and began to pull out of the dive. It took everything I had to get the plane out of the dive without pulling off the tail. As it happened, I was so low that, when the plane finished mushing toward the water at the bottom of the dive, I could see the reeds and bullrushes along the shore waving in front of me! I calculated that my low point in pulling out of the dive was less than 50 feet above the water. I never mentioned it to the commander, either; nor did I question the wisdom of his entering a dive without using the dive brakes.

CHAPTER 19

"GROUP GROPE"

by Daniel L. Polino

Part of our training in Air Group 152 consisted of night-time group takeoff and rendezvous, and finally return to base for a systematic breakup and landing.

The important thing was that this simulated operations aboard an aircraft carrier and required perfect timing and extreme efficiency, since over 80 aircraft were involved and fuel capacity and time in the air were critical.

The air group consisted of four squadrons: Torpedo bombers (TBF's), dive bombers (SB2C's), fighter bombers (F4U-4's), and fighters (F6F's), basically in order of maximum flight time capability, with the F6F's having the shortest flight duration.

There were many hazards associated with this air group operation in the dark. Any time you rendezvous aircraft at night, you have the chance of misjudging your spacing during join up, either overrunning or turning into your partner with disastrous results. Also, since the basic formation consisted of four aircraft, called a division, and with 80 aircraft in the air in one big formation, you actually had 20 independent formations, each attempting to maintain position. It's called togetherness.

To the ground-based observer, the formation looked like a big Christmas tree. Each plane had a red light on the left side wingtip, a green light on the opposite, and a white light near the tail. As the formation turned through the night sky, to the ground observer the colors appeared to change from red to white and to green depending on the direction of turn. Visibility from the pilot's viewpoint was dependent on the phase of the moon, and whether or not the stars and ground lights could be seen. An overcast condition created the situation for a real "group grope." We were all groping around in the dark trying not to run into each other.

The most difficult part of this operation was the return to base. This required systematic breakups by four-plane divisions, and landing on the darkened runway. When the sea fog decided to move over the field, as it often did unannounced, all hell broke loose. Since some of the aircraft were getting low on fuel, nobody wanted to take a wave off. The result was that aircraft would land alongside of each other, and often too close behind each other; anything to get back on the deck.

While I was enjoying my convalescence from a hernia operation at the Quonset Navy I hospital, I was joined by one of my squadron mates who had the misfortune to be returning from a group grope when the fog came in. Immediately upon landing his F4U-4 Corsair, another pilot landed his F6F Hellcat behind him and ran up his tail, cutting off slices of fuselage right up to the back of his seat. His head lacerations eventually healed, but he had a terrible headache. The armor plating behind the seat probably saved his life. Glad wasn't there for that one!

CHAPTER 20

"AIR-GROUND SUPPORT"

by Daniel L. Polino

While stationed at NAS Groton, Connecticut, our air group was assigned to air-ground support training at Ayre, Massachusetts. The U.S. Army maintained a large base where air units could attack pill-boxes, tanks, and other targets, guided by requests from the ground-based units. It simulated the type of combat situation that was common in conventional warfare, where ground forces that were held up by enemy installations would ask for help from their aircraft support.

As in the typical air group exercises, when you're trying to get 80 aircraft into the air for a strike, you always have those with the longest flight capability, in terms of elapsed time, to take off first and land last. This meant that the TBF torpedo planes took off first, followed by the SB2C Helldrivers, then the F4U Corsairs, and finally the F6F Hellcats.

We'd circle the Army reservation, waiting for requests from the ground coordinator. His counterpart in the air was the air group commander, who relayed the target information to the specific squadron and specified the number of aircraft and type of ordnance to be used in the attack.

The notable thing about the air group commander, who was also the air-borne coordinator for the operation, is that he felt that he had to be the first to take off, and that the entire air group should join up on his aircraft. Considering that he flew an F6F

Hellcat, which under our flying configuration did not carry auxiliary tanks, he really was stretching his luck, considering that some of these group exercises lasted over three hours.

I had a double dose of excitement on one specific day, part of which our group commander supplied as a result of the above-noted peculiarity. I had been selected, along with three other fighter-bomber pilots, to attack a target using our five-inch wing-launched rockets. The aircraft rockets are approximately six-feet long and the warhead is, for all intents and purposes, identical to a five-inch artillery shell.

We dove in a shallow dive toward the target in division formation; my position was off the left wing of the leader and stepped down. When he fired his rockets, we did likewise. However, during this sequence, one of his rockets misfired and dropped off its wing pylon, slewing through the air like a telegraph pole directly toward my propeller. It took some nimble rudder action to slide away from the projectile, but I managed to avoid it successfully. At the speed and altitude we were at, if it didn't explode upon hitting my prop, the rocket would certainly have wrecked the blades and caused me to crash anyway.

That wasn't the only excitement on that day. Returning with the entire air group and circling the airfield, our air group commander ran out of fuel. Remember he was the first to take off, he was flying an F6F Hellcat, and we'd been in the air at least three hours. The airfield is located on a plateau, and the end of the runways drop off to a cliff in just about all directions. You try not to undershoot or overshoot the runway when landing at Ayre!

He had time to call in for an emergency landing, then dove for the field. The tower responded by directing him to the service runway. Considering the situation he was in, he'd be lucky to reach the field, much less a particular runway! Anyway, from where I was circling I could see his plane barely making it to the flats, where he was able to land with his gear up.

The instant his plane touched the ground, one of the 100-pound bombs he was carrying split open and burned behind his aircraft much like a gasoline fire. The other bomb rolled off onto the runway, where it remained. I know it was there when I landed 20 minutes later—I nearly hit it. I'll say this for the commander: Although he was in his forties, when his F6F hit the ground and before it came to a stop, he was out of the cockpit and jumping off the wing. Maybe he saw the flames in his rear view mirrors and figured the plane was due to explode. Anyway, it was an exciting day for everyone.

CHAPTER 21

"SIMULATED NIGHT CARRIER LANDINGS"

by Daniel L. Polino

While stationed with my squadron VBF-152 at the auxiliary naval air base at Groton, Connecticut, we did quite a bit of night flying. This was always risky since the ocean fog had a way of moving in at night and covering the field in a matter of minutes. There was only one runway of any significance at Groton and it ran north and south to the ocean's edge.

Parallel to this runway, on the east side, was a long, low hill, about 500 feet high, which made a nightmare of the landing approach, especially at night. Running east-west, at the north end of the field, a railway was positioned just off the end of the runway where we would land, when the wind was coming from the south.

When shooting touch and go simulated carrier landings at night, it was the practice to follow the instructions given by the landing signal officer (LSO). He positioned himself at the landing point of the runway and, via fluorescent wands, guided us to a landing. There were usually four or five pilots in the pattern at one time and, after we had completed possibly a dozen landings, the LSO would signal us to make a final landing with a specific "stay down" signal that he'd give during the last 180 degree turn to the landing point.

As I mentioned earlier, the hill that ran parallel to the runway to the east made it necessary for us to actually fly our downwind leg on the other side of it, out of sight of the LSO. When reaching the point opposite the landing spot, it was normal to put the aircraft in a landing configuration, and start the final 180-degree let-down turn. That's when you'd pick up the LSO who would assist you by signaling if you were too high or too low, or off center.

After about the fourth circuit of the field, while coming through the final 180-degree turn, and at a fairly low altitude, I noticed a large water tower alongside the railway. I noticed it because it was illuminated by my starboard wing running light—and that was the only illumination; there were no lights on the water tower itself!. I realized then that each time I made my approach, I'd had been flying within a dozen feet of the tower!

Needless to say, from then on, when in final approach and under the assistance of the LSO, I'd glance toward the water tower to make sure I was missing it, then return my attention to the LSO for any signal he might have. After about fifteen landings, and while on final approach, I was surprised by the absence of the LSO at the end of the runway—and the fact that I didn't see any other aircraft in the pattern.

What had happened was that, when the LSO had given me that distinctive, but brief, signal to make my final landing, I was busy looking for the water tower. After about three attempts to get me down and STAY down, he went home, somewhat disturbed by his inability to get me to land. After depending on the LSO's assistance when making those practice carrier landings, I was a little nervous doing it alone, but did manage to land safely. I taxied the plane to the line, parked and secured it, and went back to my BOQ for a rest. I never did hear from that disgruntled LSO.

CHAPTER 22

"UNCAGE GYROS"

by Daniel L. Polino

During my tour aboard an escort carrier in the North Atlantic, we had some exciting night flying incidents. While qualifying for night carrier landings aboard our escort carrier, we seemed to experience continuously overcast nights with no moon or stars to help us. I don't ever remember night flying out there when I could see the moon or stars—and with the German submarines around, we had little in the way of lights aboard the carrier.

Often times, we'd be awakened from our sleep and told it was our turn to fly. This meant that you'd dress and get on deck to wait for the next Corsair to land. When a plane came in with a pilot who had made his quota of night landings, the change of pilots was very fast, in order to minimize the number of wave-offs by pilots already circling the carrier.

A plane would land, catch a wire, and come to a stop. The deck crew would disengage the hook, pull the plane back, and put chocks around the wheels. The pilot would leave the engine at idle, disengage his oxygen mask, earphones, parachute, and straps, and climb out. The replacement pilot would leap into the plane, connect the above-mentioned equipment, set up the plane for take-off, and go! This occurred in about a minute! One of my acquaintances had the misfortune to take off like that, forgetting to set his flaps for the occasion. He barely got into the air.

When I flew off the escort carrier at night, it was generally a fully instrument flight. Before leaving the deck, I set my gyrocompass and my altimeter to zero. In that way, as long as the carrier didn't change course, I'd be able to find my way back. A typical flight when shooting night landings under the conditions described above, went like this: Since I couldn't see the deck clearly, I'd align the aircraft with the faint white line in the middle of the deck—what I could see of it. I'd set my gyrocompass and altimeter as described above, go through the takeoff checklist, and go. Without looking outside the cockpit, I took off strictly on instruments, since I couldn't see the island anyway.

The gyrocompass (which read zero), the altimeter (which read zero until I left the deck), and the artificial horizon, were my guides for a safe takeoff. It worked fine; you'd fly straight ahead on zero degrees, climb to your altitude, proceed forward and make your left turn to a downwind direction, and look for the carrier. It would be coming toward you from the left side and parallel to your course. All you'd see would be the faint red light on the mainmast. Once you found it, you continued on your 180-degree course and, as you passed it, you started a gradual turn to the left, easing off on the throttle and setting up your controls for landing. After the many, many hours of simulated field carrier landings, it was relatively easy to transpose this technique into a real carrier operation.

It worked fine for me, and the many preparatory hours in the Corsair paid off one night. As described earlier, sometimes switching planes with a pilot who had completed his flight in the middle of the night could be a hurried thing. As usual I leaped into the Corsair, attached myself to the respective cords, tubes, belts, etc., went through my check list for takeoff, set my gyro instruments at zero, and took off down the deck—not looking outside the cockpit and depending completely on my gyrocompass to stay in the middle of the deck. I climbed the aircraft to altitude and began my left turn into the pattern. It was at that point that I noticed the gyrocompass remained at zero!

Apparently, in my hurry to clear the deck, I'd forgotten to uncage my gyrocompass. The only thing that kept me from going to left (overboard) or the right (into the island structure) during takeoff was the "feel" of the aircraft I'd acquired from hundreds of hours of flying, about 800 landings and takeoffs, and my trim-tab settings. I must have had some special help on that one. Anyway, it was my special secret, and you can be sure that "Uncage gyros" became a written part of my checkoff list for takeoff thereafter.

CHAPTER 23

"CARRIER LANDINGS"

by Daniel L. Polino

Landing the F4U Corsair aboard an escort carrier (CVE) during the daytime is a fairly interesting, if not hazardous, proposition; doing it at night under wartime conditions (no lights) is very exciting.

Early in 1945, we found ourselves on an escort carrier in the North Atlantic, and what was unusual was that apparently I was the only aviator aboard who had never gone through carrier qualification on the Wolverine. Usually, after graduation at NAS Corpus Christi, or Pensacola, the pilots were sent to the Wolverine, a tanker located on Lake Michigan which had been converted to serve as a fresh water aircraft carrier for qualifying new pilots in carrier landings. When I arrived there during the fall of 1944, due to a series of blizzards that are typical of the area, a large backlog of pilots was standing by, waiting to qualify. There was no way that I'd be expected to sit around under those conditions so, after Operational Training at Cecil Field in Florida, I was sent on to join Air Group 152 at NAS Gross Isle, Michigan.

After several months of training with fighter-bomber squadron VBF-152, we ended up aboard an escort carrier in the North Atlantic. Since the German submarines were out in force, we operated at night with a minimum of lights. My first daylight carrier landings were well observed by my squadron mates because they knew I'd never landed aboard a carrier before, and the conditions found in the North Atlantic were much more severe than those on Lake Michigan. Luckily, my landings were uneventful and routine, until one day when we had a nervous landing signal officer (LSO) on deck.

There were four or five of us in the traffic pattern that day practicing takeoffs and landings when, for some unforeseen reason, two aircraft in a row, after hooking onto the arresting cable, angled off to one side and clipped the metal edge of the deck with their propellers, nearly going over the side. The LSO had apparently given them a signal to "cut" their power and land while they were at a slight angle to the centerline of the deck. The result was that, after landing and hooking onto a cable, instead of slowing in a straight line down the deck, the aircraft rolled at an angle and ran over the side.

So, when it was my turn to land, the LSO had become pretty nervous and uncertain. As I made my turn into the groove, the LSO continued to give me a "roger" with the signal paddles, which means that everything is fine, all the way. Normally, you get the "cut" signal before you get to the LSO's position on the corner of the deck. Since I was passing him, still receiving the "roger", I anticipated the "cut" signal and began to ease off throttle and settle in, only to see out of the corner of my eye a belated "waveoff" signal!

Now, a "wave off" is mandatory, as it could mean a fouled deck or such, and under no conditions do you ignore it. Fortunately, one of my early instructors taught me to set my aircraft trim-tab controls for carrier landings to the takeoff configuration when coming in for a landing. The reason is that, in case of a sudden application of power while in a near-stall position, the aircraft won't flip over due to the tremendous torque of the 2000-horsepower engine.

This saved me from a sure crash as I quickly added power. I was able to keep the plane under reasonable control as I skimmed over the cables, hanging by the prop, and heading straight for the ship's island superstructure. I'll never forget the sight of the Navy photographer busily engaged in snapping pictures of what looked like an impending crash, while he was hanging off the superstructure as I headed straight for him.

Luckily, my aircraft had enough power and forward speed to carry me around the island to the surprise of the photographer who thought that he was going to get a first-class crash photo. Later, after I'd landed aboard the carrier, the LSO apologized for the late waveoff, with the explanation that, considering the two previous crashes, he wasn't sure of my alignment with the deck. His indecision nearly got me killed.

As you may know, the LSO stands on the extreme stern of the flight deck on the port side with a canvas screen to protect him from the wind. Beneath his position is placed a large net. In case it looks like an aircraft is going to crash into his position, he can dive into the net. It takes some nerve to do that, however, because you're about 50- or 60-feet above the water near the ship's churning propeller. Anyway, the next day during landing practice, I had the opportunity to get him to try out the safety net. My landing was too close for his comfort, so he chose the net. It wasn't on purpose, of course, but you could call it poetic justice.

Landing aboard an escort carrier at night in the North Atlantic during wartime is something else. Due to the constant danger from German submarines we maintained a minimum of lights. As I recall, there was a red light on the mainmast and a row of lights along the inner part of the edge of the flight deck. The flight deck lights had metal deflectors alongside to keep the light emission to about a 15- or 20-degree angle shining aft. This allowed you to see the outline of the deck when coming in over the stern for a landing, while preventing surface craft from seeing the landing lights.

The lights also provided the pilot with a perspective of the approach, such as whether you're too high or too low.

Upon returning from a night flight, assuming you could find the carrier, you'd look for the dim, red light on the mainmast; and if you knew the ship's course, you'd line up with it. Since all Navy planes use a left-hand or counter-clockwise traffic pattern around carriers as well as airfields, you could line up your gyro compass with the carrier's direction of movement and carry out your approach and landing using the compass.

There was an old saying that you could tell a pilot's marital situation by how he flies. The single guys fly low and slow when coming in for a landing—all conditions are marginal. The married pilots want everything in their favor; they keep a margin of safety by flying a little faster with a little more altitude for safety's sake.

Being single, while making my usual low and slow approach to the fantail for a night landing, I suddenly found myself flying directly over, or through, the superstructure of an escort destroyer which was apparently too close astern of the carrier. Again, the tremendous power of the Corsair's engine, when called upon, enabled me to climb over the ship's masts, awakening everyone aboard the destroyer. I was about 50-feet below the prescribed altitude but managed to climb during the remainder of my 180-degree turn toward the flight deck to the correct altitude and position. The LSO must have had some strange thoughts when watching my approach, and looking down to see my aircraft climbing up to deck level.

Carrier landings can be exciting and dangerous. I might mention in retrospect that during a two-week checkout cruise aboard an escort carrier in the North Atlantic, our squadron went out with about 21 Corsairs in flying condition for night carrier qualification and returned with about three of them still flyable. That will give you an idea of the risks involved. The Corsair was probably the hottest fighter aircraft the Navy had at the time, and an escort carrier is not a very large or stable landing field.

CHAPTER 24

"MANUFACTURER'S SUPPORT"

by Daniel L. Polino

When stationed with Air Group 152 at NAS Wildwood, New Jersey, we were delivered brand new F4U-4 Corsairs. Indoctrination and ground school for new models was conducted by a test pilot from Chance-Vought, manufacturer of the aircraft. This new model had a bit more power, a four-bladed prop, and a rearranged cockpit that included a full deck beneath the pilot's seat.

One of the points made by the instructor test pilot during the briefing was that, in case of any emergency that we might experience during the ensuing period of familiarization flights, he would be available to help us. I can remember talking to him after the class session, and his repeated promise to assist us in case of an emergency.

It was on the same morning following his lecture that I took off with part of the squadron to do some dive-bombing practice in the target area. After my second dive, I noticed a malfunction in my tail-wheel system; the panel indicator showed the tail-wheel was down when my landing gear handle was in the wheels-up position. As I suspected, the hydraulic system that controls the landing gear, flaps, brakes, etc., had failed and lost its hydraulic fluid.

Initially, I tried lowering the main landing gear, including the use of the emergency wobble pump—to no avail. At that point, I notified the control tower of my problem and, in a short time, was surprised to find the Chance-Vought representative alongside my aircraft in another Corsair. With his advice, I tried a series of actions to lower the wheels, and when all failed, he instructed me to use the emergency carbon dioxide backup system. A C02 bottle located in the cockpit deck activated by manually opening an integral valve, forcing C02 through the hydraulic system, and lowering the main landing gear.

When I opened the valve, there was a loud hissing sound and the cockpit filled with vapor; my first thought was that the system leaked. With the landing gear lever in the down position, however, it really did work; and my indicators showed the gear to be down. There was no sure way to prove that they were locked, however; so that still left a thought in the back of my mind that we could still have a problem.

Being familiar with the emergency preparations in cases of problems with landing aircraft at Wildwood, I knew that a staff of surgeons and nurses were scrubbed up and gowned, waiting to perform emergency surgery in the base hospital if the landing was unsuccessful. That didn't make me feel any better when making my approach to the field.

The approach to the landing was made at a slightly higher air speed to take into account that my flaps were inoperative. When landing a Corsair under these conditions you prefer to make an Air Force type of landing where you land on the main gear, rather than a typical Navy three-point, full-stall landing. It was probably my best wheels-landing; I couldn't even feel the impact of the aircraft on the runway—just greased it in.

The run out down the runway was another thing, however. There was barely enough hydraulic fluid left in the accumulators for the main gear brakes to keep the aircraft going straight down the runway; it tended to veer to the left. The result was that I burned out the right wheel disc brake, but saved the plane—and myself. Of significance to me was the fact that Chance-Vought's representative was there when he was needed, and made good his promise of that morning. Now that's what I called manufacturer's support!

CHAPTER 25

"THE JOHNSON BOYS"

by Daniel L. Polino

Fighter-Bomber Squadron 152 (VBF-152), with Cmdr. John Devane, Jr., as our skipper, was rather unique in that it was composed at two groups, segregated by age. There were the new pilots like myself, fresh out of operational training, inexperienced, and eager to join the fleet in the Pacific. Then, there were the old timers, some in their late twenties, who had at least one tour of duty under their belts and a chestful of decorations attesting to the fact that they were survivors. We always tried to show the older pilots that we could fly.

One of our senior pilots had been shot down in the Philippines and had spent some time with the guerrillas before being brought out by submarine to give information to our forces for MacArthur's invasion. Our executive officer had flown with "Blackburn's Irregulars" (VF-17), a high-profile unit. All were survivors, both because of their unique abilities and pure luck. They certainly were men to emulate.

There were three different pilots with the last name of "Johnson" in our squadron. One of them had nearly been shot down by a Japanese Zero that had gotten onto his tail during a dive bombing attack, and was spraying him with bullets. Luckily, one of his squadron mates was behind the Zero and eliminated him. Johnson was very nervous—but brave. He endured countless hours of dangerous flying.

During the eight months we were together, we lost two of the three Johnsons. The above-mentioned pilot was the only survivor among the Johnsons. The day the Japanese surrendered, this remaining member of the Johnson clan went to the skipper, took off his wings, and asked to be taken off of flight duty, which was granted. He had certainly been through the mill, and probably felt that there was a curse on his name. He had plenty of guts, though, to stick it out as long as he did.

One of the Johnsons who lost his life while I was with the squadron was the division leader who I flew wing on. While on a familiarization flight one day with a new Corsair model, I was flying over the general area near Groton, Connecticut, watching a tail-chase by some of our squadron pilots a few thousand feet below me. Tail-chasing is sort of a follow-the-leader game. The leader performs aerobatics and the rest duplicate his maneuvers, each following closely behind in a kind of string. It was during a slow roll that one of the planes fell out of the roll and the pilot attempted to recover with a split-S maneuver. There wasn't enough altitude and, as a result, he hit the ground at about a 45-degree angle, the pilot being killed in the impact and the resultant fireball.

I knew it was one of our boys but didn't know which one. I gingerly coaxed my plane back to the base and found upon landing that it was my division leader. It took guts for the remaining Johnson to continue daily flying with our squadron until the end of hostilities.

CHAPTER 26

"CROSS-COUNTRY WITH THE NAVY RESERVES"

by Daniel L. Polino

After the big war (WWII), many of us in the Niagara frontier area expressed an interest in maintaining an active unit in Naval aviation. We finally did get the Naval Air Station (NAS) Niagara Falls commissioned. I was one of the original six officers.

For awhile, I was attached to a fighter squadron, flying F4U Corsairs. During this period, most of us who flew with the Navy at that station were either going to school under the G.I.Bill or had a job. The air station had a full-time crew that maintained operations for us "weekend warrior" reservists. They were called station-keepers. I took a two-week vacation from my place of employment, Bell Aircraft, in order to serve my annual two-week active duty with the Navy.

One of the station-keepers put together a cross-country trip from Niagara Falls, New York, to Jacksonville, Florida, via Norfolk, Virginia, and Charleston, South Carolina. His idea was to lead us to Charleston, where he lived, and enjoy a few days there while we completed the flight to Jacksonville. We'd then return without him.

We navigated Air Force style, following railroads and such to Washington,D.C., flying at a low altitude over the Washington monument (that would in today's environment get you shot down), and finally reached Norfolk where we refueled. When we reached Charleston, landed, and then prepared to take off, I found that I'd blown my tail wheel tire. We sent the other four aircraft on to Jacksonville, found a new tail wheel tire, and changed it ourselves (the station keeper and myself). I spent a fun evening in Charleston, sightseeing and eating some of the fine Southern seafood. The following day, the four pilots returned from Jacksonville, and I joined them for a return trip to Norfolk

After refueling at Norfolk, and with storm warnings alerting us of the poor conditions between Norfolk and Niagara Falls, we decided to try to by-pass the weather by flying toward Philadelphia, then to Niagara Falls. I was now the senior officer and responsible for the flight of five aircraft. Following takeoff, we immediately ran into severe thunderstorms with heavy rain.

It was our good fortune to find the Naval Air Station at Chincoteague, Virginia, where we landed; any port in a storm. There we were, civilians at a spit-and-polish U.S.Navy base, where everyone salutes anything that moves. We were out of money, out of clean clothes, and anxious to get back to Niagara Falls because our two-week leave of absence from work or school was about over. We washed our socks and underwear each day in the lavatory sink, and managed to subsist on our remaining funds. When you're at Chincoteague, you may as well be on the moon.

Each morning, we'd load our suitcases into our planes and wait for operations to let us take off. The overcast remained with us for several days. You could see blue sky through the clouds, but they wouldn't let us go. Each evening, we'd take our suitcases back to the room. After several days of reasoning, and a lot of pleading, the operations officer let us take off. It was a Saturday.

We proceeded on a straight line to Niagara Falls, crossing over continuous ridges of the Appalachian Mountains. The overcast hung down to the peaks, allowing us to pass through the unobscured saddles. Luck was with us; we were able to reach our destination in a few hours.

The weather was so cold at Niagara Falls, that since we wore only our undershorts under our green gabardine flight suits, we nearly froze before getting to the hanger after parking the planes. All in all, it was an interesting flight. I've since found out that Chincoteague is famous for its native horses and annual roundup; nothing else.

CHAPTER 27

"TOWING THE TARGET BANNER"

by Daniel L. Polino

After WWII, the Naval Air Reserve was activated at the Niagara Falls Airport. The airport was shared by the Navy, Army Air Force, National Guard, and Bell Aircraft. A typical once-a-month duty weekend would consist of flight training sessions including high altitude gunnery over Lake Ontario, or anti-submarine training exercises. Whenever we went out on a gunnery flight, one of our squadron would take the duty of flying the target banner.

The banner was a rectangular piece of woven nylon approximately 50-feet in length and 5-feet wide, attached to a 1500-foot line, and towed behind one of our F4U Corsairs. To maintain the banner in a vertical position, a bar with a heavy steel ball was affixed to the juncture of the cable to the banner. The trick during takeoff was to face the aircraft upwind at the end of the service runway, lay out the 1500-foot line with the banner attached ahead of the aircraft, and attempt to lift the banner without dragging or tearing it.

This involved a typical short runway takeoff. To accomplish this, you set your flaps to full-down, held your brakes, stick held full-back, and ran the engine to full power. When you released the brakes and eased forward on the stick, the tail would rise to align the plane's body with the horizontal and allowed quick acceleration to get you airborne. Within a couple of hundred feet, the stick would be pulled back to enable the aircraft to literally jump off the runway. The wheels would then be retracted, the rpm's reduced to climb configuration, and a steep climb begun, gaining a maximum amount of altitude before the tow line snatched the banner off of the runway, undamaged.

Returning from a flight, the banner would be dropped on the grass alongside the service runway at a signal from the tower. Bullet holes in the fabric could be matched with the aircraft on the flight by color coding applied to the belts of ammunition by the ordnance personnel prior to flight.

If you towed the target on a particular flight, it meant that you had to take off before the squadron, and land after all other aircraft had landed. This meant stretching your fuel supply, since some of these flights lasted several hours. When it happened to be my turn to tow the target, I took precautions during the flight to minimize my fuel usage rate. Since the towing speed is usually about 150 knots, I leaned out the fuel mixture and flew at reduced engine speed (rpm's). Of course, this makes the engine run cool, for which you close the cowl flaps.

After a typical, long flight period, I was flying back to the Niagara Falls Naval Air Station, towing the banner, with a single F4U escort to preclude aircraft from running into the tow line, when I started a downwind turn around the airport. Alerting the tower of my intentions to parallel the service runway and drop the ball, line, and banner on the next pass, I prepared for my cross-wind turn at 1000-feet of altitude. At that point, possibly because the engine was too cold or I was low on fuel, or something else was wrong, the engine began to cough and sputter, threatening to quit. It would run properly only 50% of the time!

A quick message to the control tower, stating that I was making an emergency landing with the banner still attached, brought a routine response—"Use runway 270." Of course, my plan was to use the first and nearest runway. The airfield boundaries were fenced in with an 8-foot-high chain link fence, and there was no time to release the towed target!

My emergency landing involved landing with 1500-feet of

line and the previously described banner configuration trailing behind. Fortunately, the banner cleared the fence; otherwise, either the tail would have been pulled off the aircraft, or the fence would have been torn down. Fortunately, there was minimum damage to the banner, enabling the scoring of hits by the other pilots. I never did find out why the aircraft's engine missed so badly prior to and during the landing.

Engine malfunctions you learned to take in stride. If you had to scrub a flight for every discrepancy discovered while taxiing out for takeoff, you'd not get to fly very often. Sometimes the problems were found after you'd become airborne. Anyway, it turned out to be a very exciting landing.

VBF-152 at Brown Field, California, Sept. 1945. Dan Polino is in the third row, fifth from the left. (Photo courtesy Dan Polino)

Corsair coming aboard just after the "cut." (Photo courtesy Dan Polino)

PART III

"Letters from the Bird Barge"

by Owen W. Dykema

To Owen Dykema, free lance writing and publishing represents one of four parallel careers in a busy life:

(1) Husband, father and grandfather: With Enid, his wife of 48 years, Mr. Dykema has two daughters and three grandchildren.
(2) Navy carrier fighter pilot: Flew the last of the big prop planes and the first of the jets, including a combat tour in Korea. After an additional 30 years in the Active Reserves he retired as a Navy Captain, and now resides with his wife, Enid, in Roseburg, Oregon.
(3) Research Engineer (BS/MS ME): Spent 14 civilian years as a "rocket scientist", helping to develop rocket engines such as those that powered man to the moon and that power the space shuttle today. He then spent another 20 years developing environmental technology, in particular a process to burn coal without polluting the atmosphere.
(4) Writer / publisher: Writing credits include 20 technical papers; some 70 OpEd pieces and letters in Los Angeles and Oregon newspapers and in Time, Atlantic Monthly and Scientific American; four pieces in a history of the Flying Mid-

shipmen (FM); a gold medal in the writing contest of the 1996 Texas Senior Games; and two self-published books, "Letters From the Bird Barge" and "Legacy of the War Orphans: How We Lost World War II." He is currently editing and publishing a CD version of the FM history (about300,000 words), and a history of the bomber crews in the Southern Oregon Warbirds Association.

Owen is president of "An Association of Writers (AAW)" and of Dykema Publishing Co., in Roseburg, Oregon.

"LETTERS FROM THE BIRD BARGE"

Book back-cover copy:

In the summer of 1952 I flew 47 combat missions in the Korean war in the F4U-4 Corsair fighter-bomber off the carrier Princeton. As a brand new husband and father at the time, I wrote my wife a letter almost every day, and she saved them all. In 1992 we reopened them. It was immediately obvious that none of the scary stuff had found its way into the letters. So, from 40 year-old memories, I added that in. The result is a unique combination of quiet husband-wife romance interspersed with the roaring excitement and adventure of carrier-combat flying.

And you are there . . . late at night in the little cone of light at the writing desk in the Junior Officers' Bunkroom, late at night in a lonely cockpit over the dark Pacific qualifying in night carrier landings, and in the adrenaline-charged atmosphere of yellow-orange fireballs and angry black bursts of anti-aircraft fire in the midst of a dive bombing run.

The following two chapters are excerpts from this book, "Letters From the Birdbarge" It is available in paperback only (214 pages, 27 pictures, 2 cartoons) from Dykema Publishing Co., 3264 W.Normandy, Roseburg, OR 97470-2406.

September 1997

ISBN: 0-9660705-0-X

Price (incl. S&H): $12.00

CHAPTER 28

"Deck Crash & Night Landings"

by Owen W. Dykema

March 23, 1952

Part of the letter Owen wrote to his wife:

> Sunday is a day of rest, right? We all worked harder today than any since we came aboard. Especially my radio men. Poor guys have a tremendous amount of work to get done, and only a few days to do it. They were all up to midnight last night, and two of them worked all night. They're really putting out. They sure look tired and PO'd. Bunch of good guys.
>
> I got in the movies today. About thirty of us came up to the flight deck on the deck-edge elevator and ran past the grinding cameras of Monarch Films, Inc. Then we had a mass start of planes, and four of us, including yours truly, taxied past the cameras again, spreading our wings. I was in 203B, with my goggles up, and no crash helmet. In case you ever see a movie of the Princeton, CV-37, I'm in there somewhere. That's the plane I crashed the barrier in, by the way. I suppose they'll include some action shots when we start operating, day after tomorrow.

Here's what Owen DIDN'T Write his wife:

3-BLEC

My "barrier crash" on the carrier qualification ("CarQual") cruise was one of those scary things that I didn't tell you about. It was not really so scary in itself, it was sorta funny, actually. However, I was pretty concerned about what it meant for my future on the cruise. How did I manage to get myself in all those messes?

It seems that one of the major problems with the Corsair as a carrier plane was the distance the cockpit sat back from the nose (from the propeller, engine, etc.). It was not originally designed that way but, to make room for an additional internal fuel tank, ahead of the cockpit, Chance-Vought moved the cockpit aft more than a foot. Made for a long distance from the cockpit to the nose, hence nicknames like "hose-nose" and "hog". As a result, when you were taxiing and, especially, when you were flying slowly with full flaps, gear and hook down, coming aboard the ship, you couldn't see anything directly in front of you.

The U.S. Navy, or at least Air Group 19, in all its wisdom, taught a landing approach to a straight deck carrier in which you would first fly down the length of the ship (port side), opposite to its direction of travel, then make a steady 180-degree turn to the left, to end up wings level, lined up with the centerline of the ship and about 30-yards or so astern, ready (in a second or two) for a "cut", and landing aboard. The trouble with that approach, especially, and perhaps uniquely, in a Corsair, was that during those seconds when you were approaching and then flying up the ship's wake, ready for a cut, you couldn't see the ship at all. All you could see, down in the lower left corner of your windshield, was the Landing Signal Officer (LSO). During that short period of time you were totally dependent on him, for your life.

So, there I was, flying alongside the ship, about 100-feet over the waves and about 1000-feet (about 2 blocks) out from the ship, landing gear and those big barn-door flaps full down and locked, hook down, and slowed down to about 95 knots (about 110 mph). At that speed the plane was pretty cocked up, with that long nose slanting upward, and the plane was sorta mushing along through the air, just a few knots above stall. No problem. With the plane fully trimmed for this condition it just floated along. I was holding an accurate altitude above the water by making sure a certain spot near the top of the ship's mast was exactly on the horizon.

On this day, with the cockpit open and the weather bright and sunny, all seemed right with the world. What an adventurous and thoroughly enjoyable life! What would I rather have been doing? About halfway down the side of the ship I rolled in about 30-degrees of bank and started my (approved) standard rate turn to the left, toward the ship. To maintain altitude while in this bank I had to add a little power and roll in a little more nose-up trim. Still holding that spot on the mast right on the horizon, still no problem. I was mushing along toward the ship in a gentle turn, beginning to approach the ship's course, lining up on the ship's centerline and beginning to fly up its wake.

At that point I flew into the blind spot. I could no longer see the ship nor, especially, my spot on the ship's mast. Furthermore, right at this time, right in the blind spot, I had to roll my wings level, take off some power and lower the nose just a tad to maintain my altitude. All this without any accurate visual reference to tell me if I had done it all exactly the right amount, or not. The LSO was clearly in sight and he seemed satisfied with the result this time. He gave me no corrective signals and then—there it was—a cut. I chopped power back to idle, shifted my gaze away from the LSO to the ship, and began to lower my nose for a landing.

But wait a minute, something was very wrong. I was not 20-30 feet in the air like I was supposed to be, like I usually was; I was almost on the deck! And what's worse, I was going like a bat out of hell. What a shock, what had happened to that calm, pleasant day and flight? Suddenly I was quite sure that I would rather be doing almost anything else but this. I put some back pressure on the stick, trying to get the tail and the hook down where it could catch an arresting wire. Unfortunately, I was going too fast so the plane rose higher in the air, taking the hook with it. Okay, a little forward pressure on the stick. The plane came down nicely but now the tail and the hook were up in the air, where the hook couldn't find a wire. So there I went, porpoising down the deck, trying to get lower, feeling for the deck and a wire.

All of a sudden the crash horn sounded. It took me a long several milliseconds to recognize: "My God, that's for me!!! I'm still in the air and the ship is predicting that I will end up in a crash!" This whole episode, from cut to stopped, took less than five seconds but I had plenty of time to consider the uniqueness of the situation. Actually, it didn't occur to me to wonder if I would be hurt or if I would even survive the crash. Rather I felt a kind of perverse thrill because, by God, there I was flying up the deck with the crash horns wailing and every eye on me, the very center of attention!

I clearly remember noting that I was drifting a little to the left, and had time to lower my right wing and start flying back toward the centerline. Finally, the plane began to run out of airspeed. Both the plane and the tail eased down lower, and I was nearly on the deck. On the other hand the barrier, a set of very strong, ugly-looking steel cables stretched across the deck, about cockpit-high, was still coming at me at about 70 mph. Finally, there it was, I caught a wire.

It turned out to be the very last wire before the barrier (#14). With a violent jerk I stopped. The barrier was just in front of the prop, scraping on it on its downward sweep. I cut the engine to minimize further prop damage. It turned out the prop just had to be filed smooth again, and some dents in the engine cowling hammered out, but otherwise, there was no significant damage. After all, there was the old adage of straight deck carrier pilots: "There's them that have hit the barrier and there's them that are going to." (Navy pilots never were all that great at the King's English.) That day I joined the group of those that have hit it, but only lightly.

But the LSO was howling mad! It turned out that just as I rolled out of my turn, and lost sight of the ship, I dropped my nose just a hair, and started coming downhill, picking up speed as I came. The LSO claimed that he could do nothing but cut me. He feared that, had he given me a "wave-off," I would have poured on the power and raised my nose to go around, thereby lowering my tail and hook. In that attitude the hook might well have latched onto a wire, with me still up in the air. Up in the air, full power on, and the hook attached firmly to a wire! In that event I probably would have flipped over and destroyed me and a lot of other things and people in the process. So he claimed I forced him to take the best of two bad choices, to give me a cut. One might have asked (but not too much by an Ensign of a Full Lieutenant) why he didn't give me any "get your nose up" signals right when I started coming down hill. Never got an answer for that.

Now the LSO was an experienced pilot. As a full lieutenant he must have had some 5-10 years of experience, probably including some time during WWII, and he had many more flight hours and carrier landings than I. However, we were talking about my life here and I wasn't at all anxious to trust his experience, and blindly obey, when he seemed so clearly to be in error. If he is wrong it is me who might die, not him.

I was far more interested in resolving the basic problem, the one that might get me fully into the barrier one day. This was the problem of this particular carrier approach, inherently containing this moment of blindness, right at the time when you had to make all of those simultaneous attitude and power changes, with little or no chance to observe and correct for any errors. I pleaded with him to let me come over the fantail (the rear of the flight deck) and take the cut while still in my bank. That way I would not have to make any attitude or power changes before the cut, my left wing would be down and I would still be turning so I would have a clear look at both the ship and the LSO, and I'd still be able to see my "altitude" spot on the ship's mast. After taking a cut, however, it would then be up to me not only to lower my nose and land but simultaneously to roll my wings level and seek out the ship's centerline to land on. That still seemed infinitely more "doable" than the blind spot thing.

After much arguing the LSO finally said: "Well, all right, I'm going to let you kill yourself. You do it the way you want to but I refuse to take any responsibility for the results." So thereafter I would wait a little longer before starting my turn toward the ship, turn a little tighter, until there was just a little water showing between the fantail and the nose of my plane, then ease off on the bank a little, trim up for that turn, and just sit there right up to the cut. I felt like thumbing my nose at the LSO as I went by. He could clearly see me, and I him. Rolling out and landing on the centerline, as I thought, turned out to be a piece of cake. I was the only one aboard ship that made this kind of pass so everyone knew when I was coming aboard. Now wouldn't I have been embarrassed if I had crashed and killed myself?

[I found several decades later that both the British and French Navies found the same problem with their Corsairs and, after numerous trials, adopted my approach (that I discovered independently). I also found that my approach had been standard procedure in at least one other Korean War vintage Air Group with Corsairs.]

NIGHT CARRIER QUALIFICATION

The part that really worried me, though, was how the LSO was going to behave toward me now, for the rest of the cruise. Case in point, my night qualifications later on that same CarQual cruise. For some reason, which I can't recall, all pilots in our squadron had already qualified aboard ship at night except me. One night I was scheduled to complete my night landings, and I was the only one on the schedule! Flight operations that night were conducted solely for the benefit of yours truly.

So there I found myself, all alone up on the port catapult (cat) in the inky blackness shortly after midnight. The Corsairs we were flying were not particularly intended to fly at night. We had a small group of Corsair night fighters but they had flame dampers on the engine exhausts, radar on one wing tip, and pilots specially trained for night flying. It's true that if you stay in the dark long enough you become "night adapted," and you can see considerably better than you would imagine possible. That is, if your eyes don't see any appreciable white light (red light is okay).

So, that night the launch director signaled that they were ready to shoot me off. I got all set for the powerful forward acceleration, head back against the rest, fingers of my left hand hooked around a grip on the throttle, right elbow tucked in against my body. The night out there looked black as the ace of spades. There was just enough light fog in the air that there was little or no discernible horizon. Inside the cockpit, the familiar soft red lights and the glowing engine and flight instruments were reassuring. At the very least I could fly instruments if I couldn't orient myself visually.

When I ran the engine up to full power the engine exhaust stacks began shooting out very hot gases which, late at night, without flame dampers, were bright yellow-orange flames. The flames themselves were not very visible to me but they reflected off the propeller, turning it into (for a night-adapted guy) a bright yellow-orange disc, right in front of me, right where I had to see! But, I have to say, there was absolutely no way that I would ever have attempted to chicken out in a situation like that, in front of God and the entire ship. So, scared to death, I gave the "GO" signal to the launch director.

Suddenly there was the powerful jerk and push of the cat and, in just over two seconds, I found myself thrown off the bow at 100 mph, out over the black, unforgiving ocean. As always happened in cat shots, at least with me, the shot pushed my goggles up on my head such that the rubber padding along the bottom ended up directly across my eyes. For the first few seconds I had to let the airplane fly itself while I pulled my goggles down, so I could see. This was followed by a flurry of activity to get the gear started up, adjust power and rpm, and get set up to fly. Finally I had a chance to look outside.

There was nothing out there! I had never seen such total, black blackness. And as I suspected, I could not distinguish any lighter shades of black that might serve as horizon. The only safe thing to do was to fall back on instruments, and that brought a rush of relief. Nevertheless, I was flying instruments in the black of night only 100-feet or so above the ocean. And I was only 23 years old, just a few years out of a small town in Illinois, only two years of college, and much too young and inexperienced to die.

So after flying upwind into the ink bottle for a while I started a gradual left turn back to the ship. If all went as I was briefed, there would be a destroyer, with at least minimal running lights on, positioned correctly on the port beam (left side) of the carrier such that if I flew directly over him I would be in the correct position to start my turn toward the ship. Somehow I felt certain there would be another typical screwup and the destroyer wouldn't be there. Part way through the turn, looking over my left shoulder, I did in fact see a dim collection of lights back there that looked like the destroyer. As the turn went on another set came into view that looked like the carrier. Saints preserve us, it looked like all was going according to plan!

Now, with at least some lights available for orientation I could at least partly get off the "gauges" and start flying visually, and get ready for landing. Gear, hook and flaps down, and slow it up. Prop pitch full forward (had to be ready to GO if I had to), and cock that big nose up in the air (the plane's, not mine). All seemed set, the plane was mushing along downwind, all trimmed up and stable, at about 110-mph. There went the destroyer underneath. About halfway down the length of the ship I started the turn in. Now just lay the aircraft nose right alongside the stern of the ship and let the ship just pull me around to a landing. Trim tabs, power, attitude and altitude all set and stable.

Approaching the ship, I could see the LSO (my new-found "friend", the LSO) on the port side, with the "black light" illuminating the fluorescent strips on his arms and legs, and on his "paddles." He was giving me nothing but a "Roger" signal ("you are looking good, no corrections necessary, just keep coming as you are"). Now I was coming in close—now almost right over the LSO—the Roger remained fixed all the way—ready for the cut— . . . . what's this, a wave-off?! I eased on full power, lifted the nose a little, stayed in the left turn till clear of the ship, then rolled back to the right, to fly up the port side, going upwind again.

Driving by the ship, except for the dim "dustpan" lights casting some illumination across the deck, and a few tiny running lights on the mast, it looked "closed for the night". No signs of life or activity were apparent, the flight deck looked deserted.

So then I was in the upwind inky blackness again, on instruments. What was wrong with that pass? The LSO showed nothing but a Roger all the way. All the way to a wave-off, that is. As best I could judge, everything was as it should have been, I should have gotten a cut. So, then, what did I do wrong? It was true that I had never landed aboard ship at night, was there something I was missing, or forgetting? Was the hook not down? Wouldn't the LSO have advised me of that? (then it occurred to me—was this some little game the LSO was playing, to "get back at me"? In fact, was it he that had gotten me up here, all alone? . . . Nah-h-h.)

So, back down the port side I went, this time double checking everything in readiness: gear, hook, flaps, power, prop—everything was okay. Checked again—yes, all was as it should be. Over the destroyer again—started the turn—smoothly floating around the turn, like I was on a wire—there was the LSO—there was the Roger again—looking good—the Roger was holding—now I was coming up on the ship—over the LSO—WAVE-OFF!!

Honk it around, up the port side, the ship still looks deserted. I was indeed the Flying Dutchman—doomed forever to circle this deserted carrier, making endless perfect Roger passes to endless perfect wave-offs. There was absolutely no radio traffic, I seemed totally alone in this black night, in this part of the world.

Back up into the ink, back around. Down the port side, over the destroyer (a deserted destroyer as part of the timeless scenario?). Really well trimmed up this time, the plane was just drifting around the turn, the ship pulling us into the proper time and place with a perfect attitude, altitude, airspeed—everything. There was the Roger again, all the way, there was the LSO, and there was—ANOTHER WAVE-OFF!!

Now I was starting to panic. There I was all alone out over the Pacific ocean, several hundred miles from any land, in a land plane, in the inky blackness of night, and I couldn't seem to get aboard the ship. I was beginning to think that I might eventually have to set it in that cold, black forbidding water and take a chance on being picked up before I froze to death. And there was no one to help me—it was me or not at all. There was not even any radio traffic to put me in contact with another human being. An even bigger specter, if I really started to panic I could so reduce my confidence and capability that I would not be able to get aboard no matter what. Keep calm, I told myself, keep your confidence up.

Around the race track again, another perfect pass, another perfect wave-off. Then another. Never any signal from the LSO but that perpetual, perfect Roger. Was I really that good or had the LSO simply decided never again to give me any kind of indication of problems in my approach? He did say ". . . you don't need me . . . I'm going to let you kill yourself . . . "! It had been about an hour, now, and still no sign that I would ever get aboard. The Flying Dutchman went endlessly on. The engine roared, we circled around, and there was no evidence of life anywhere.

Suddenly, the radio crackled into life . . . "Uh-h 203, this is the LSO. Just thought I would let you know that your passes are good, the deck has been foul (not in condition to accept landings), so I thought you could get in some good practice. We should be able to begin landings on your next pass." Suddenly I wanted to rise up out of my seat and strangle someone (i.e., that LSO). He had let me go all that time without advising me of the problem, he had let me sweat out here all by myself!! Didn't he have any compassion for a first time young Ensign out there alone in the black of night? Was this my punishment? I'll bet my face was red as a beet!! In this man's Navy, however, with the LSO a full lieutenant and yours truly the lowest of the low, an "Enswine", you can bet I wouldn't comment to him at all. Besides that, he still often had my very life in his hands. I just had to depend on myself alone to fly the pass and get aboard as much as possible independent of the LSO.

So, down the port side again—over the destroyer—into the turn—there was the "LSO(B)", holding his usual Roger—and this time there was the cut. Perfect landing, right on the centerline, about number 4 wire. Taxi forward onto the port cat—another shot off into the ink (raise the goggles and regain control). Around again to another perfect landing. Same with four more, and I was night qualified. All night long I never got any signals whatsoever from the LSOB except the Roger and the cut. After I parked the plane and shut down the engine, the ship seemed unearthly quiet, with just the rush of the wind over the deck. It was true, there was almost no one in sight on the flight deck. The LSOB had long since gone below. Do you suppose there is a lonely Dutch Ensign still out there, making endless passes at the ship and taking endless wave-offs? It was way past time to go below and hit the sack.

* * *

(Several days later) coming back from Barber's Point, Oahu, we had 16 planes flying in a somewhat unwieldy but nice-looking, tight formation, on the way out to the ship. The skipper led us too close to a thunderhead, and before he could turn us further out of the way my sideof the formation flew into the thunderhead. It was BLACK and bumpy in there, and thick enough to make it hard to see your leader, and to keep your wing position on him. It was a little bit antsy for a while but then we broke back into the open. All was well except with ENS Harvey Winfrey. Although he was still flying his tight wing position on his division leader, he was now UPSIDE DOWN!

Somehow (we have never been able to figure it), Harv kept sight of his leader in the cloud, and maintained his relative position, but got inverted! Harv himself had no idea how it happened. It was the subject of much speculation for days. Really weird things like that seemed to keep happening to Harvey. In our original "CarQuals" he bounced his plane clear OVER the barrier, landed and skidded to a stop just short of the bow. The only "non-arrested" landing the PRINCETON ever had.

CHAPTER 29

"PARALLEL PARKING IN THE F4U"

by Owen W. Dykema

After yet another agonizingly long and uncomfortable flight, and the usual tenseness of landing aboard, I was finally on the deck, forward of the barrier, and could begin to relax. Following the signals of various taxi directors I was taxiing slowly toward a parking spot near the forward end of the island. I could hardly wait to shut down and get out of that airplane.

With my wings folded I couldn't see anything or anyone forward of the cockpit, of course, but that's what the taxi director was there for, exactly in that little triangle formed by the fuselage, the horizontal (gull) part of the wing and the outer wing folded back (about 120 degrees) over the fuselage. Everything was proceeding normally, in just a few more minutes I would be down in the wardroom relaxing over a cup of hot coffee.

Therefore, it came as a great, startling shock to hear that 140 db crash horn start blasting right in my ear—BONG—BONG—etc. For an instant I mentally checked over my own status, and everything seemed okay—except—I noticed that the taxi director was looking not at me but down the deck. His eyes were large as dinner plates but he was still waving me on, into the parking spot. Finally it came through to me, someone was headed up the deck for the barrier—for me! And there was no telling whether he was in position to be stopped by the barrier or if he was going to come on over it, right into me. Judging by the size of the director's eyes the guy was coming to get us! I could just feel him crawling up the back of my plane!

All of this thinking was taking place in about a second—from relaxed casualness, at the end of a long, tough flight to full alert, with a full charge of adrenaline flowing. During that second I continued to roll forward , and it was exactly that critical one second that got me in almost irrecoverable trouble. I think the director came out of his initial shock first, just a hair before I did. He had time to recognize that he had let me get too close, that the time for stopping was long since past, and got off an emergency stop signal, and I began reacting. The brake pedals still had to travel forward about an inch before they began to clamp on the brakes. During that travel I started crunching the plane ahead. Things simply happened too fast and, as luck would have it, at exactly the worst possible time. Considered this string of events (counting seconds by the old "thousand one" method) from the start of the crash horn: "BONG—Thousand—BONG—one—BONG—thousand . . . CRASH, BANG, etc. . . . BONG—two."

I ended up about two feet up the plane ahead of me. My prop cut off the hook (a 15 or so pound chunk of steel) and threw it into the director's back. Pieces of metal were flying everywhere. I damaged the prop and cowling on my plane and threw shrapnel through the two planes parked next to me. Four planes were downed in one fell swoop! People up on vulture's roost, watching the recovery, saw the hook pull out of the landing plane, and watched that plane proceed sedately up the deck and eventually into the barrier—no big deal. But they weren't prepared for the hail of metal parts and fragments that then blew down the deck from my crash.

So, who's fault, and what to do to prevent it happening again? I think it really was just a freak accident, with necessary events and timing that may never have gotten together before and very likely will never get together again. If I had been a more experienced and "seasoned" carrier pilot, I might already have fixed in my mind that nothing works normally while the crash horn is sounding, and might have stopped instantly, as soon as it began sounding.

Instead I'm a 23-year old kid, just starting out on my first extended deployment aboard ship. The taxi director may have been even younger and less experienced than I. Wars are fought by people like us.

Nevertheless, I was very embarrassed and shook up over the accident. I may even have been (may still be) unconsciously blotting out some of the events. My worst regret, besides hurting the deck crew, was that, as far as I can remember, I didn't even stop to see if the director had been hurt. I didn't see him get hurt, so I wasn't aware of it at the time, but I should have investigated. My memory is that my Division Leader was walking back from his plane and saw it happen. When I climbed down from my plane he said "I saw it all and it was not your fault. Come on, let's go below and have a cup of coffee." And I went. I will live with that for the rest of my life.

The next day the guys had a mock ceremony in the Ready Room where they awarded me the "Hero Medal of North Korea" (cut from a tin can lid).Together with my previous barrier engagement I had damaged five American planes. That made me an "Ace" for the other side. And I had yet to start flying combat!

ENS Dykema ready to mount his trusty Corsair and go forth to do battle.

ENS Harvey Winfrey bounces over the barrier cables, landing safely near the bow. (USS Princeton 1952 cruise)

PART IV

William "Country" Landreth, CDR USN (Ret.)

I interviewed Bill "Country" Landreth extensively in 1993, and reduced the interview to a story of about 7000 words. It is mostly about his time with VF-17 "Jolly Rogers"—also known as "The Skull and Crossbones Squadron"—the first Navy squadron to fly the F4U Corsair in combat in World War II.

Originally from Tilden, Nebraska, "Country" retired with his wife, Ginger, in Camarillo, California.

CHAPTER 30

"F4U CORSAIR COMBAT PILOT"

by Fred "Crash" Blechman

William L."Country" Landreth flew F4U Corsairs with the VF-17 "Jolly Rogers" and the VF-10 "Grim Reapers" (later VBF-10) during the 1943-1945 period of World War II in the Pacific Theatre. He was the youngest original pilot in VF-17 and is credited with three air combat victories before being captured by the Japanese in March 1945 and serving six months as a prisoner-of-war until repatriation.

"Country" spent another 24 years in the Navy in various assignments, including Commanding Officer of VS-37, a carrier-based anti-submarine squadron. From VS-37 he went on staff to Commander Carrier Division 17, then to the Naval Research Lab, then became the Pt. Mugu Liasion Officer to Vandenberg Air Force Base. He was the founding director of a degree-granting program at Naval Air Station Pt. Mugu, California, and helped to start degree-granting programs on two other military bases.

Bill Landreth is authorized to wear the following ribbons and medals: Distinguished Flying Cross with Gold Star; Air Medal with 4 Gold Stars; Purple Heart; Navy Unit Commendation; Prisoner of War Medal (finally authorized in 1991!); and numerous theatre ribbons. He received a Bachelor Degree at the age of 47 from Chapman College in Orange, California, just before retiring as a Commander in 1969.

After retiring, Bill spent five years in the educational field and received his Master of Arts Degree in Education in 1975, and a Master of Science Degree in Counseling in 1976, both from Cal-Lutheran University in Thousand Oaks, California.

Presently living in Camarillo, California with his wife, Ginger (and celebrating their 54th anniversary in 1998), Bill has been self-employed in a specialized form of insurance for financial institutions for almost the last 25 years.

I interviewed Bill just a week before he and Ginger were headed for the 1993 VF-17 Jolly Rogers 50th anniversary reunion in Denton, Texas. Here's his story:

Fred: Did you have an interest in flying prior to going into the Navy?

Country: Oh, yeah. Sure. That's all I ever wanted to do. When I was about eight years old I used to get banana crates from stores, drag them home, knock them apart, nail them up, and go around making noises like an airplane. I built balsa wood and tissue paper airplanes, but they would never fly. When I got through I had at least three-quarters of a pound of glue in the darn thing and it would fly like a brick!

I used to chase the barnstormers. Everytime one of those planes would fly over, I'd run a mile across a plowed field to go over to that airplane—but they never took me up for a ride.

However, after one year of college, I got a private pilot's license in the CPT (Civilian Pilot's Training) Program in 1940, flying a 50-horsepower Cub. This was in the summertime between semesters. I was the first one to solo from a group of about fifteen. The following summer I went to secondary CPT in Sioux Falls, South Dakota, flying a Waco biplane, doing aerobatics. This was identical to the C-Stage of Navy flight training.

Fred: Why Navy flight training?

Country: I'm originally from the little town of Tilden, Nebraska, population of 1100 people—now 990 people. When I was in high school I looked around and saw grocery stores, gasoline stations, feed and seed stores, hardware stores and tools that farmers need—but no industry. What kind of a future was there in Tilden? When you got out of high school you had to figure out some way to make a wage and make a good foundation for a family. I saw people who were there all their lives and were pumping gasoline.

I thought, well, I need to do something that's more difficult than that. And from all my reading, it seemed the most selective, the most difficult to get into, the best training—was carrier aviation in the Navy. It was the most demanding, the toughest. So my next thought was, gee, that's a pretty high level goal to try for. Then I thought, if I start at the bottom it'll take me a lifetime to get up to the middle, so what I'll try to do is start at the top and then if necessary I'll fall back down to the place where Peter's Principle will take me—the highest level of incompetence!

Fred: How old were you when you went into the Navy?

Country: The Navy required two years of college before flight training. I had the required two years of college, but I was too young—you had to be 20. So I went back for a fifth semester. On the third of December I was 20, Pearl Harbor was on the seventh of December, and I was in the Navy on the twenty-sixth of December—the day after Christmas 1941.

Fred: Tell us something about Navy flight training.

Country: I reported to duty at Fairfax Field in Kansas City for "E-Base" (Elimination Base) training. They gave us ten to twelve hours of dual flight instruction before solo. If you didn't solo by then, you were washed out—and could apply for the Army Air Corps!

I soloed and was sent to New Orleans, where we flew nothing at all; we went to ground school. We were waiting to be fed into Pensacola on one end, or Corpus Christi on the other side. Which ever way you went depended on who had the space for you next. I went to Corpus Christi, Texas. They had patrol planes, dive bomber training, fighters—but none of that happened until the last stage. My total flight training lasted for nine months. For instruments we had Vultees; we sat in the back seats under a hood. Our fighter training at the tail end was in SNJs. The last SNJ hop I had was gunnery on a towed sleeve.

I got my wings and commission at Corpus in October of 1942, and went to Opa-Laka in Miami, where I flew Brewster Buffalos—F2As—for so-called pre-operational training. I'm glad I never had to fly that thing aboard a carrier! By December I was in Norfolk attached to VF-41, where we checked out in F4F Wildcats and waited for the commissioning of VF-17.

The Wildcat was a great airplane for 1938, but this was almost 1943. It was underpowered. Flying it out of Norfolk, which had short runways at that time with power lines at the end of the field, you'd crank on the power and the Curtiss Electric propeller would howl and growl and make a lot of noise and the airplane would just sit there and say, "Who? Me?" Finally you'd get up to some kind of flying speed, hoist back on the darn thing, and hope it would come off. You had to crank the wheels up. From the side our guys like Hal Jackson and Danny Cunningham, fellas that were kind of short, you could see their heads bobbing like this when they were down on the crank, and the airplane was bobbing like a roller coaster. All the while they're trying to get the wheels up they would look up and see those wires getting closer and closer!

Fred: I understand you were one of the original members of VF-17 when it was formed.

Country: You're talking to one of the plank owners of VF-17. On the first of January, 1943, those that were sober enough to do so, reported for muster at eight o'clock in the morning. The skipper mercifully read his orders commissioning the squadron and said, "Muster will be tomorrow morning at eight o'clock. Dismissed."

We flew the birdcages (the original F4U-1 cockpit canopy) up until we got deployed to the Pacific in combat, where we flew the 1A with a semi-bubble canopy. I'll tell you, with that birdcage you couldn't see diddly-squat. But I'm eternally grateful we flew Corsairs; that was the best combat aircraft alive in the world at the time.

Fred: VF-17's skipper was LCDR Tom Blackburn, and some called the squadron "Blackburn's Irregulars." Was this a rag-tag group of castoff pilots from other squadrons?

Country: Yeah, but I don't know but that was overdone a little bit. In some cases I think that was true. We had some pretty tough guys to handle. In the case of myself and several others that came from Corpus via Miami, we were raw Ensigns with no experience, and we were sent to that squadrom because it was being commissioned at the time we were available.

You have to be aware that everybody that came out of Corpus Christi wanted to be a fighter pilot. Some are selected and some are not selected. As it turns out, Bob Keiffer and I set a record in the Training Command the last day of our training in the number of hits on the sleeve. We both had the same identical number of holes in that sleeve, and both broke all the records they had before in the SNJ. I feel then, and now do, that may have had a lot to do with us going to the premier fighter squadron in the Navy.

VF-17 was the first successful carrier-based Corsair squadron—and there wasn't any fighter pilot that didn't want to fly the Corsair, the best airplane in the world. Among other things, it meant you had the best chance of survival because of performance and what-not.

Fred: Tell us about your skipper, Tom Blackburn. Is it true he was a strict disciplinarian on duty, but pretty-much anything goes off-duty?

Country: I think "anything goes" is out of whack. He demanded top performance, he demanded you to be there on time ready to fly, he expected you to be an expert formation flyer, he didn't tolerate any lack of discipline in the air. He was inclined to be lenient about things on the ground until you crossed that invisible line. He didn't have a lot of spit and polish attitude. We had our inspections, we went through all the other drills. He tried to make us Navy people as well as fighters.

He certainly was a party man himself. I learned early that you don't go ashore with the skipper's group and try to stay up with him, because it was a useless task. He'd be the last one standing, and he'd be the guy that would be there first the next morning.

He was a young-looking slender fella'. In fact, more than one Ensign, reporting for duty, went to the squadron ready room, and (seeing Blackburn) said, "Hey, fella'. Where's the skipper? I'm reporting for duty." Blackburn would say, "Yeah? You're looking at 'im!"

Fred: Did you have some carrier work before going into combat?

Country: We had quite a bit of time on the carrier "Bunker Hill" (CV-17) before we were taken off just before going into combat. The supply line was all full of parts for Grumman airplanes in the fleet, and the Marines were operating Corsairs in the Solomons and they had all the parts down there. So when we got to Pearl Harbor in Hawaii, the Navy decided they couldn't support us.

I can tell you about landing a Corsair aboard a ship. You've done it yourself, and if you got it around and landed safely on board, you done did it! It was a pretty good feeling of accomplishment when you got back home to the carrier again. During the approach you always wondered about whether you ought to open your cowl flaps or not, because if you open them you can't see, and if you close them the engine heats up. I had the advantage of being tall enough—I'm six-four—so I could kind of stick my neck up and see around that cowl.

You know, though, speaking of carrier landings, a lot of these modern Navy aviators don't quite understand what it was like to take a cut where there was no alternative. There were the barriers and a pack of airplanes in front that could blow up and kill you. After World War II, I went through almost a ten year period where I was carrier-based flying ASW (anti-sub warfare) Grumman S2Fs from straight and angled decks—over 350 total carrier landings. Angled-deck landings are a piece-of-cake compared to what we used to do. Most aviators who were raised on angled decks just don't know what they missed!

Fred: So you went into combat land-based? Where?

Country: If you had a map here, it would be simple to show you. Moving up the Solomon Islands chain from Guadalcanal, our next leap was to New Georgia. Nearby Munda, where there were mostly torpedo and dive bombers, was a fighter strip called Ondongo ("Place of Death".) That's where we were based from October 27 until December 2, 1943.

Among other things, we were flying cover over the Marine amphibious landings at Empress August Bay in Bougainville. That's where a lot of our airplane victories occurred—to other people, not me—because over the beachhead Japanese planes were flying in from Rabaul trying to interrupt that landing. That's the five-day period where we shot down 60 airplanes or thereabouts.

Fred: How were the living conditions at Ondongo?

Country: Powdered eggs. Have you ever had any powdered eggs? They look like something you don't want to eat until you taste them—then you're sure you don't want to eat them! Powdered milk, and Spam. I guess the potatoes were real. We had Quonset huts, with cadet-type upper and lower bunks. When it rained they were not quite waterproof and we called it "Mud Plaza."

That's the main place we saw "Washing Machine Charlie" (Japanese bomber.) One lone airplane—bzzzz, bzzzz, bzzzz, bzzzz—he would come over every night regular as clockwork and drop bombs randomly. He never did any damage that I know of, but was very distracting when you tried to sleep. . . They'd throw the searchlights on him, shoot at him, but as far as I know never put a hole in him. This was before our night fighters were available.

Fred: I understand you lost a good friend on your first combat mission.

Country: The way we lost Johnny Keith was one of those silly, terrible accidents. Like most accidents, fatalities or casualties, they are the apex of several things going wrong together.

On my first combat flight, beginning the late afternoon of November 1, 1943, I was flying Johnny Keith's wing. We went out on a fighter sweep over Kahili Airstrip on Bougainville.

The flight leader invented an idea on the way back. We were going near the Shortland Islands on our way from Bougainville back to Ondongo. This guy says we'll spread out and go low on the water and come in from the north. There was a harbor there. The theory was that there would be some landing craft or barges bringing supplies in hidden amongst the trees, or anchored out in this little lagoon. We won't come from the direction of our base, we'll be coming from the other direction. We'd get low on the water, spread out, and pop out over that ridge right into the harbor, and shoot up everything that was a good target, a decent target of any kind whatsoever.

Boy, they were waiting for us! I'll tell you, they saw us from I don't know how far out, but when we popped out over that ridge I've never seen anything like it before or since. At about 200 feet or 300 feet, our altitude, 40mm explosions like a flat black roof! I'm telling you, they must have had those guns set on instantaneous—automatic. Every gun that was around the whole harbor just fired straight up in the air—all these black bursts of 40mm fire. It didn't look like 5-inch stuff. It was thick and intense—and we were right in the middle of it with all our airplanes—eight, as I recall, but it could have been twelve.

I left Johnny for a moment; I thought I saw something along the shoreline. It looked suspiciously like a barge or something, and I whipped around to focus on that and I strafed the entire shore, including this suspicious-looking area. I turned back on the line, joined back up on Johnny, and while I'd been gone, he got a piece of this anti-aircraft fire. He was streaming white smoke—which is engine oil vapor. So I knew right then that Johnny was not going home that day.

I flew right on his wing—no further than that window over there—and his airplane was working all right. We were in communication by hand signals and radio, but it was just a matter of time before he would run out of oil and his engine would seize. We withdrew from the island, and Johnny knew he was going to have to make a water landing. He was in trouble. As his airplane slowed down I stayed right with him, right down to the water as he made a good water landing.

Now here's the way you get killed. Procedures! You're going to make a water landing? Ditch the hood! Two red handles, in, down—it's gone. At least get the canopy locked back, with your elbows, like this.

But Johnny wasn't a very big guy and not a very strong guy, and he didn't do emergency procedure. You've got to get the gosh-darn canopy off so it can't hurt you, can't damage you, can't trap you. I'm sitting there watching this. He would push it back and it would slide back—but it wouldn't lock. Then he'd fiddle with his (microphone) cord, cinch up his shoulder straps and seat belt and the canopy would slip forward, then he'd push it back again, then he'd do something else, then push it back—but it never locked. Because it wasn't locked, it slid forward and closed when he hit the water!

The airplane, of course, floats for a little while nose down, but it went down pretty quick. He had most of his ammo and quite a bit of gas left, so it was a heavy airplane. He didn't have much time, and when he hit the water, now he's locked in. So he sort of half-way panicked and he opened his canopy and tried to get out—but you've got your parachute, your boat, your survival gear hanging on you, trying to get out of a narrow opening. He decided he's not gonna make it that way. So he unsnaps his parachute and boat, he gets out through the narrow opening, stands on the wing root and decides to reach back into the airplane to pull the stuff out—and the airplane sank. He was left with his Mae West lifejacket in shark-infested waters, right off the coast of a Japanese-occupied island—with nighttime falling. I stayed with him, and two other guys climbed to altitude to call for a Dumbo (PBY rescue plane)—all the while knowing that open-sea landing at night is not what PBYs do. As slow as they are, they could not get there before nighttime.

I stayed until it was dark. The next morning we went out pre-dawn looking for Johnny—a whole bunch of airplanes flying all over the area. We never found him. So I lost my best friend the first day in combat. It was a little difficult to handle that.

The flight leader got in trouble with the higher-ups for what they called "jousting with anti-aircraft." They said that's a non-profit proposition. That island was bypassed. All they had was ammunition on that island; they didn't have any food. So we flew by and gave them a nice target. That was dumbness!

I knew Johnny all the way through flight training and our squadron training and out to the Pacific. Johnny was a slight individual and one of the finest gentlemen that ever walked around in a pair of shoes. He didn't have a mean bone in his body, and didn't have a very aggressive attitude.

I always kind of thought he was probably too polite and too generous of heart, and would have been a good patrol plane pilot! He wouldn't go swimming over at the swimming pool at the Officers' Club with you, and it took me six months to find out that he thought he was too slight of build and he had knobby knees and he didn't want anybody to see him in a bathing suit! That kind of tender-hearted guy is usually not mean enough to be a good fighter pilot.

Fred: Did you have any air victories while flying from Ondongo?

Country: No. Except for the hotshots who were up in front all the time, it was hard to break through and get into the action. You could hear their machine guns firing when they had the mike button down and yelling, "Watch out! He's on your tail!" or "Get him offa' me!" while we were bending the throttle over the quadrant trying to get there! During the first half of my combat experience in the South Pacific with VF-17, those of us in the rear divisions got there just after the fight was over. I remember when Andy Jagger got his first victory after a long dry spell, and a photographer took a picture of him in elation. That picture was published worldwide, and is now on display in the Smithsonian.

Fred: What other action did you take part in while flying from Ondongo?

Country: We shot up some boats at Buka and Bonis at the north end of Bougainville. We had some exciting hops. We had a big pre-dawn launch one time on station above Empress Augusta Bay. Radar was directing us and we were in and out of the clouds at night flying four planes in formation when I saw the outline of a Tony (Japanese fighter) going by in the opposite direction about 500 feet above us. He didn't see us.

I had a decision to make; to pull back on the stick and try to chase him through clouds and darkness and find him and kill him—but that would break up our formation. By the time I decided I better go get him, he was gone and it was too late. All these things went through my mind. There's three other airplanes besides mine milling around in the clouds with this Tony; I wondered if I would have shot one of my own guys down. All you have is just the exhaust to shoot on.

Fred: VF-17 left Ondongo on December 2 for a two-week leave in Australia, and returned to Espiritu Santo awaiting reassignment. On January 24 the squadron flew to its next base. Tell us about that.

Country: By then the Seebees got two airstrips, Piva Yoke and Piva Uncle, right near the beach of Empress Augusta Bay, (Torokina, Bougainville.) We used Piva Yoke as a base to fly to Rabaul every day. The Marines just took enough area around there for the airfields. They didn't want the whole island; the Japanese were all over the island.

We relieved VMF-214—Pappy Boyington's "Black Sheep" squadron—on station. We flew in, they flew out, just about two weeks after Boyington had been shot down and captured.

Fred: Did you ever get any bullets in your airplane?

Country: I never got a bullet in my airplane. I'm gonna give you a sales talk now. Wally Schub, his wing man, and I and my wingman, went through all the combat—and remember, now, there's sixty or seventy Zeros sitting over the target waiting for us when we got there—we went through all the scraps, chasing airplanes away from the bombers—that was our job, and we never lost a bomber to air combat—and with all that falderal and chasing planes away, and covering each other, and keeping ourselves from getting killed, none of us got a bullet in our airplane in that division. And we came back from each mission as a unit. We left, flew as a unit, we fought as a unit, and came back as a unit—which is not so of a great many of our people. When we returned from a mission, we were four planes in formation.

Fred: Your first personal air victory was on February 5, 1944. How did that happen?

Country: As I recall, we were at high altitude over the bombers. They would go in, drop their bombs, and we would cover their bombing run and their retreat. The danger was for the bombers to be overtaken by Japanese fighters as they were going out of the target area to their rendezvous point. We were following their rear.

I was leading the second section in Wally Schub's division, with Clyde Dunn flying on my wing. In the turn, as you're leaving, it's a little difficult to keep from dragging behind, so I was a little behind and covering the bombers' withdrawal. Wally was out toward the bombers and I was catching up with him. He was turning, sort of, toward me a little bit. I don't think he saw—he's always claimed that he did—a Zero coming in on him, beneath him, coming up from below. I remember hollering at him, "Wally, quick, turn toward me! Turn more!" I had to get him to bring this guy around. And he did. He tightened it up a little bit, and as he did that, well, then the Zero tried to turn—and that put him right where I wanted him. So I had to pull up on him, and with the long nose of the Corsair, you've got to pull up for lead on a deflection shot, so the target was clear out of sight under the nose. I squeezed off a good solid burst, eased the pressure on my stick, and here came this ball of flame out from under my nose! I guess that first burst had just nailed him. Then I pulled up and gave him another squirt just for good luck. It was a 60-degree or better deflection shot. I was coming in on the top of him, and he was in a bank, and I was matching his turn.

That moment of elation was when I knew I got my first victory. It was astonishing to experience after all that dead time—all that dry run time I'd had when everybody else in the squadron was shooting down airplanes but me. I really did have a soft spot about that business of not being able to catch up with the victories of the other pilots.

Fred: Did you even wonder how you'd feel if you shot one down?

Country: Oh, sure. I always wondered. Yeah. There's no use trying, I found out, to pre-experience. Because it's not going to be the way you think it is. Each individual probably has had a different experience. . .

Fred: Tell us about your February 7 flight.

Country: I got one airplane before the bombers had done their job, and then when we came back I got another airplane. Same flight, but after we had finished the bombing run and come back, I got a second airplane that day. But here's the deal; I could have gotten several more!

Before we took off that day, I had a standard practice of having my plane captain count the bullets—the 50 caliber bullets—that came out of my guns as I charged them with the hydraulic chargers. I wanted to know not only that my guns were loaded, and had a round in the chamber, but I wanted to know if my hydraulic chargers were working—all six of them.

So on this particular flight I had arranged with my plane captain to signal me when he was ready for me to charge my guns. I charged them—and nothing happened. At most, one or two of them charged out of the six. So I motioned to the kid that I was trying again. The hydraulic gun chargers were not working properly.

The protocol was, if your guns are not working, and your chargers are not working, you miss the flight. Now this was going to be my opportunity—remember the "dry-run kid"—and I knew this special deal was laid on. They're not leaving me behind. Not Uncle Bill. Not this time. I had a feeling this would be the day.

So I called him up to the cockpit and I said, "Get your butt out there on those guns and manually load and charge all six guns, and leave them that way, so they're ready to fire." He did that and I took off.

My wingman, Clyde Dunn, followed me around all during the flight. After the bombers had gone, the skipper and a couple of other people stayed up and flew high cover while the rest of us went down and attacked the Zeros going home.

I ran up on probably half-a-dozen airplanes with my Corsair. I followed a Zero through about three split-Ss that day. I never did get him, because when he saw my guns fire, he did a split-S, right underneath me, like that. I'd turn over and pull, but you couldn't pull that Corsair through like a Zero. I'm pulling 4 Gs, and my airplane is shaking and stalling, and

I'm firing, and the guns are throwing the ammunition all over the sky—and I'm not touching him. I had no means by which I could catch that guy. He finally just flew away unharmed.

Meanwhile, my guns are quitting, one by one. With all the Gs, and all the sometimes-negative Gs, I was finding airplanes, I was fighting them, I was getting them dead to rights, I'd press the trigger—and I'd get two guns! I was so disconcerted by that time, I'm gonna' fix this. I hit the hydraulic gun charger so I could get my guns back. I wound up with one gun. My left outside gun was all I wound up with, and I was in the middle of the best dogfight I ever had in my life!

Over Lakunai Airfield, Rabaul, I came up on the tail of one Zero, after I'd already shot down one, and I'm telling you, he was just a dead duck. He was either so lackadaisical about paying attention, or he was so scared he was frozen, because he was going straight back to the base like that. I flew right behind him with Clyde Dunn on my wing, and I started shooting. Everytime I'd squeeze the trigger the gun recoil would kick the airplane off line because it's the way outboard gun, and none on the other side. So I said, okay, then I'll fire the gun and then I'll push the right rudder and I'll see-saw back and forth and I'll cut him in two. Never touched him.

I started out a different system then—I'm still trying to get this guy. So, I said, I'll just aim to the right of him, and then as the gun kicked me across it'll get him. Didn't touch him. He's straight and level. No problem. I'm right up behind him. Couldn't get a bullet on him.

All this time Clyde Dunn is following me through all these machinations and all these movements and so on, and I figured, well, surely Clyde's going to shoot any minute. That guy never fired a shot the whole day!

We had never specifically discussed this. We should have. I feel it's a failing of mine that I didn't say, "Now, Clyde, when you see my guns go, I don't care where you are or what you're doing or where you're pointed—unless it's at me—you squeeze and you fire!" We'd have shot down about four more airplanes that day if he'd have done that; his guns were all working.

Fred: You later lost your wingman, Clyde Dunn. How did that happen?

Country: My wingman lost his life because, as I described earlier, "a unit flying together and coming back together" was violated. Here's the second way a man gets killed, loses his life: overeagerness to get into the fray in a situation that was not the way it should have been.

The next day (after the February 7 flight) a pilot that had come up and flown one hop, and caught a 20mm, turned in his wings. The skipper sent him down. Clyde Dunn heard about it, and went and volunteered to go on that flight. I heard about his volunteering and I said, "Clyde, you're my wingman. You belong in this division. If they need extra pilots, we'll all four take the hop. I'll talk to Wally and I know that he'll go. The people that leave together, fight together, come back together. Don't go with a bunch of strangers you haven't been flying with. You don't know if they'll protect you or not, and you don't know how they're going to fly or what they're going to do."

This is the way you survive. You get so you can read that guy's mind. You know when he dips his left wing what you're supposed to do. You go with some other folks, alright, you might survive, but you might not. Clyde went with them, against my wishes, against my pleading—I didn't just ask him, I pleaded with him. I said, "Clyde, tell them the whole division will go if they need planes." He went, and didn't come back. . .

He was so eager to go and redeem himself for not having fired a shot during all that combat, he wanted so badly to get into the fray, that he was willing to take the chance that he took. Much to my regret. Clyde was one on the finest gentlemen we had in the squadron. But, there it is. You're thinking is what gets you sometimes. You think yourself into a hole like

Johnny Keith did by deciding he could get out of the airplane without his gear and reach back and get it—and the canopy should have been gone. That's the way you lose your life.

Fred: I understand you were the last VF-17 plane to leave when your tour ended.

Country: Yeah. The day we left (March 8, 1944) we were under 75mm artillery fire and there were a couple of Corsairs burning. As I was taxiing away from my revetment, my plane captain jumped on my wing and reached inside the cockpit and grabbed my first aid kit. He said the plane captain of the revetment next to me had been seriously wounded. I got out of there in pretty rapid order after that.

Fred: After VF-17's combat tour, what happened?

Country: We showed up in San Francisco on the first of April, 1944 and the squadron broke up on April 10. After some leave we hung around until we got a new set of orders. The orders I got were to Green Cove Springs, Florida, near Jacksonville, for Combat Team Training flying Corsairs. This was the Summer of '44.

The new concept was instead of sending individual pilots out, they'd replace a whole division or an eight-plane combat team, the leader being a combat-experienced pilot like myself. So we went there and we flew with what we called our "nuggets" (new pilots.) We did our combat team training, and had orders to the West Coast where we would then be a replacement unit to go aboard ship in some squadron.But I got a change of orders while in Nebraska. I was ordered to return to the East Coast and report to Air Group Ten, which was forming up VF-10, the "Grim Reapers," flying Corsairs. Here's what happened. Will Rawie was the Superintendent of Aviation Training down at Green Cove Springs, and he had been on the beach and figured to get out and get his command and get into the war. I guess he had some good friends in Washington, because he scraped all combat experience out of all those combat teams that were headed for the West Coast. He kicked all of the VF-17 guys right out of their teams and sent them to the East Coast to form up in his Fighting Ten. He had a lot of talent in that squadron—although he had eviscerated those combat teams headed for the West Coast. They were left with only nuggets!

We went through a lot of training with VF-10. By that time, the wisdom of the Great Fathers up in Washington had decided to split the fighter squadrons into two—a fighter squadron (VF) and a bomber-fighter squadron (VBF.) I went with VBF-10. We had a real horse to carry those bombs with—that Corsair was a great airplane with a lot of horsepower.

Our training included firing Tiny Tims at China Lake, California. The Tiny Tim is a 500-pound warhead on about a twelve or fourteen foot long tube, which is the rocket motor. It weighed a lot—a lot of steam there—but we still could sling two under each Corsair. We qualified with Tiny Tims, and on graduating day our skipper, LCDR Will Rawie, took up two divisions—eight Corsairs. Each airplane had two Tiny Tims and eight five-inch high-velocity rockets—four on each wing. That's a total of sixteen Tiny Tims and 64 rockets. We dove the entire formation of eight planes in concert and salvoed the whole shooting match at a range target! That dust cloud was seen from many miles away, and I understand that they still talk about that salvo at China Lake—almost 50 years later.

In February of 1945 we proceeded to Alameda, California, where we were hoisted aboard the carrier "Intrepid" (CV-11), which was often called the "Decrepit" because every time it stuck its nose out of the harbor it would get another torpedo.

Fred: What about your combat experience with VBF-10?

Country: It lasted one day—the eighteenth of March, 1945. I was the first plane launched for fighter cover in the fleet that morning when we were attacking the Japanese home islands. I had the black, pre-dawn launch with my division. We had no enemy contact.

On the second flight that day, we were on a fighter sweep, looking for enemy aircraft, looking for trouble of any kind, over Shikoku. We found no air opposition, so our secondary mission was an airfield near a town called Iwajima. We came down from upstairs at high speed to strafe the field. While retreating out a sort of fiord at low altitude, and still going at high speed, I saw a speedboat going lickety-split, kicking up white water. I thought there must be somebody important in that boat or it wouldn't be going that fast; I'm gonna go get that boat.

I headed for the boat, pulling high Gs. Between me and the boat was this little island that had some round tanks on it that could have been small storage fuel tanks or what have you. The buildings were sort of handsome, with red tile roofs, so I figured I'd give them a squirt on the way by to get this boat. I gave them one burst, then shifted my attention to the speedboat, and I was just pulling down on the trigger for the second burst when that whole red tile roof just turned inside out—a big red ball of flame and black smoke. I went through that thing at 300 feet and about 400 knots. This was a high order explosion—not just fuel.

The airplane came out the other side perfectly okay—except for some small detail. But the explosion was so strong that it crushed the ninth thoracic vertebra in my back. I tested the wings to see if the Corsair was all in one piece or not. It flew fine. Then I thought, it's gonna be tough to try to land this airplane aboard a carrier with the injury I have.

There was enough pain—there was no mystery about it—that I knew my back was badly hurt. But guess what? That little round dial down there was reading zero—no oil pressure! Following emergency procedure, I successfully jettisoned my canopy as the engine seized, and made a nice dead-stick water landing. No problem; I had time to get out of the airplane and get my boat and parachute. I floated around for three days, covering myself with the tarp with the blue side up so the Japs couldn't see me. I was hoping a Dumbo could slip in and pick me up. Everybody saw me go in, and I was in radio contact—but this was in an estuary in the home islands of Japan, so you're not going in there on a Sunday afternoon drive.

About the third day, in freezing rain, with my feet frozen, double pneumonia, a broken back, and stricture of the esophagus (which I still have after 48 years), I finally got picked up by a Japanese boat.

Fred: And they nursed you back to health?

Country: They didn't nurse me. I just stayed on the floor of the prison cell and survived—as well as I could on rice three times a day. But that's another whole story. . .

P.S.: The State of Texas House of Representatives adopted resolution H.R.No.627 on April 28, 1993 to "commend the gallant men of Navy Fighting Squadron Seventeen on the occasion of their 50th anniversary reunion. . . "

In this painting, artist Jim Laurier has depicted a VF-17 mission on February 7, 1944. After safely escorting U.S. bombers in and out of Rabaul, Commanding Officer Tom Blackburn led several of his Corsairs back to the east side of Rabaul's Simpson Harbor in a low-level strafing run on Lakunai Airfield.

At the top of the painting six Jolly Roger pilots are shown, left to right: "Timmy" Gile, "Country" Landreth, "Tommy" Blackburn, "Rog" Hedrick, "Danny" Cunningham, "Ike" Kepford.

The original of this 16 by 22 inch color painting sells for $95. Jim Laurier, a fellow member of the American Society of Aviation Artists, has many other paintings, and is available for commissioned work. He may be reached at (603) 357-2051.

PART V

"TAIL END CHARLIES -Navy Combat Fighter Pilots at War's End"

by Former LTJG Roy D. "Eric" Erickson

USNR VBF-10

Eric's excellent 267-page large-format 8-1/2- by 11-inch hard-cover book was originally self-published, but then reformatted to 160-pages and republished by Turner Publishing Co., Paducah, KY, where it can be ordered for $34.95 plus $5 shipping and handling at 1-800-788-3350, or from the Turner website at www.turnerpublishingco.com . (KY residents add 6% sales tax).

There are so many good Corsair stories in this book it was hard to select the three that follow, which reflect some of the "true tales of trial and terror" of flying Corsairs near the end of WWII against Japan.

Former LTJG Roy "Eric" Erickson USNR is authorized to wear the Distinguished Flying Cross with Gold Star, Air Medal, Asiatic-Pacific Area Service Ribbon with one Star, American Area Service Ribbon, WWII Victory Medal, CBI Patch, and Expert Pistol Ribbon.

Eric attended the University of Nebraska, Pomona College in Claremont, CA, and the Art Center School in Los Angeles, CA prior to the war. He entered the Navy V-5 Program as an AvCad and upon completion of flight training received his Naval Aviator "Wings of Gold" and officer designation as Ensign in early 1944. His book details his wartime experiences.

After the war Eric attended the Chouinard Art Institute in Los Angeles, CA. He married Sammye Jean Williams, a girl from Texas, and they celebrated their 52nd wedding anniversary in 1998.

Eric was an art instructor at the University of Wisconsin, illustrated national magazines, and exhibited his paintings nationwide. In 1949 he taught school at the Chouinard Art Institute in California and started his own business, Erickson Associates, Inc., destined to become one of the top ten interior design firms in the country.

Erickson Associates has designed environments for such personalities as Richard Nixon, Robert McNamara, and Jay Chiat. The main body of work included hospitality resorts such as Sun Valley in Idaho, the Boca Raton Hotel and Club in Florida, and Cross Keys Inn in Baltimore, Maryland. He has acted as a design consultant for Westlake Village and Janss Corporation in California. Eric has lectured nationally on interior design and written many articles on color and design.

He is represented in "Who's Who in the West" and "Who's Who in Interior Design."

After raising two sons, he is now retired in Pasadena, California, and remains active with writing, painting, sculpting, and computers.

CHAPTER 31

"Carrier Landings—Day & Night"

by Roy D."Eric" Erickson

A routine carrier landing was neither overly complex or difficult as long as the pilot knew what he was doing and cooperated with the LSO, who assisted the pilot in carrier-landing by using only arm and hand signals. Although the signals were, in most cases, an indication and not an absolute order, the pilot was obliged to follow these silent directions that had been developed over many years.

It was important to develop confidence in taking direction in this unnatural fashion, and there were a few serious do's and donuts. The pilot, for example, had to trust that the LSO's judgment was sound, and resist the temptation to "chase the deck" by pumping the control stick up and down. A certain amount of wind over the flight deck was required to reduce relative motion and ease the shock of the landing on both ship and aircraft. The pilot was required to rig his approach a few knots above stall speed while the carrier increased speed to make more wind or, alternatively, slowed to reduce it. I always added five knots for my future family, and many times got reprimanded for doing so.

Landing an F4U Corsair presented a greater problem than any other type of aircraft. The visibility over its long nose was nil, and if you opened your cowl flaps it was difficult for some to even see the LSO. When the Corsair was first introduced, it was doubtful that it would ever make a good carrier aircraft. The British however, who were one of the first to put the Corsair into carrier service, had discovered that if they approached the carrier from a turn, straightening out just prior to landing, they could follow the LSO's signals and come aboard without mishap. We were soon to follow their lead. The F4U also had tricky stall characteristics, and trim tabs had to be constantly adjusted to hold the correct speed and approach. The Hellcat was a snap to land compared to the "hog."

Typically, a ship might launch one or two divisions, four or eight aircraft for CAP (Combat Air Patrol) and recover them three or four hours later. On air strike or bomber escort missions, however, carrier aircraft often operated in large groups of 40 or more from several carriers simultaneously. The flight was routinely so noisy that talking was useless. Despite the bustling activity of launching and recovering, moving aircraft was accomplished calmly and almost entirely without talking.

Most of my flight operations began with a catapult launch. I was guided by a director, taxied to the catapult and required to straddle it. A cable bridle connected the aircraft to the catapult shuttle, and a hold-back fitting was placed between the plane's tail and the flight deck. The launch officer, standing alongside the aircraft, signaled me with a rapid rotating motion of his uplifted right hand to apply power. The fitting at the tail restrained the aircraft during engine check.

If the cockpit instruments indicated everything was satisfactory, I saluted or held up my arm to the director and pressed my head back against the headrest in preparation to launch. When the launch officer was satisfied that all was ready, he knelt on one knee and made an elaborate arm motion forward that ended with two fingers pointing toward the bow. This silently dramatic signal prompted the catapult operator, stationed in the catwalk, to activate the hydraulic ram. The hold-back fitting parted and the aircraft hurled along by the catapult, was on its way.

Aircraft returning from a mission flew to their carrier and entered a left-hand orbit over the standard rectangular flight path. Formations of four aircraft passed the ship on the starboard side on the same course and flew ahead a distance of a mile or so. The division then made two 90-degree left turns to return abeam of the ship at a distance of approximately a mile. Once abeam, the formation again turned left and flew at an altitude of a few hundred feet toward the carrier, repeating the pattern until the ship was ready to recover them. Additional divisions intending to land orbited above those ready for recovery.

Landings commenced when the carriers turned into the wind and hoisted the "Charlie" flag at the yardarm, indicating to ships within visual range that the carrier was "landing aircraft, stand clear." After flying up the starboard side and ahead of the carrier, the first aircraft to land banked left, turning away from its formation. The second aircraft to land turned away from the formation about 30 seconds later to follow the first aircraft. The number three and four aircraft followed in turn.

Although I did not time the upwind turn, I based it on my "seaman's eye." It was adjusted to achieve an optimum landing interval of 34 to 35 seconds. Our interval under combat conditions was set at 15 seconds, but usually came in around 20.

As each pilot left the formation, he passed the lead to the next by patting his head and pointing, though more often than not the only signal was a casual wave. The flight leader normally landed first, gauging his turn to roll wings level as the ship steadied on a landing course. If he was early, he had to take a wave off. A correctly executed turn, however, pleased everyone.

The aircraft established itself at about a half-mile on the carrier's port beam at 150 to 200 feet, flaps and tail hook down, doing approximately 90 knots, depending on the aircraft type. The LSO picked up the approaching aircraft as it turned left toward the ship, and from this point on, I focused almost entirely on the LSO.

Stationed on a platform on the port side at the extreme aft end of the flight deck, the LSO held brightly colored cloth paddles for easy recognition. A large canvas blind protected the LSO and his assistants from the strong wind over the deck, usually about thirty knots, and comments were logged for later discussion after my landing.

The first signal after the left turn to final concerned the approaching aircraft's altitude. If at a proper height, the LSO held both arms straight out from the side of his body at shoulder height, the "roger" signal. If it was high, the LSO advised me by raising his outstretched arms slowly above his shoulders; if low, the LSO lowered his outstretched arms below shoulder height.

Next in rapid succession came signals to remind me to lower flaps and tail hook if either was not already down. Speed control signals followed. If my aircraft was slow, the LSO moved his outstretched arms toward the plane and back, a "come to me" motion. If the approaching aircraft was fast, the LSO slapped his right leg once or twice with the right hand paddle to tell me to reduce speed. In a less graceful move the LSO shook a leg at the arriving aircraft if I was flying in a skid.

The most critical signal, "turn left to align with the flight deck," came just before touchdown. Some leeway was possible in speed and altitude, but the turn to final was critical. Once commencing the left turn abeam, the aircraft approached in a continuous, descending flight path all the way to the cut position. The LSO, with arms outstretched, leaned to one side at the waist rapidly two or three times to show by the angle of his arms the amount of increase or decrease in the turn needed to line up with the deck. This motion was followed by the cut signal in which the LSO smartly passed a paddle across his throat to instruct me to close the throttle, shift my gaze to the deck and land. I then, in response, leveled my wings and straightened the aircraft to align it with the flight deck. Once at the cut position, minor differences in altitude, speed, and alignment from one landing to the next required me to be alert and fly the aircraft to the landing.

If the LSO judged everything was not set up for a safe landing, I received a wave off signal (a deliberate wave of the paddles above the LSO's head). If received, I was obliged to increase power and go around for another try. The waveoff signal was given well before the critical point in the landing sequence. In many cases, the reason for the wave off had nothing to do with my performance.

Although I was under LSO control on a carrier landing, I must fly responsibly. For example, I was not to slam the throttle wide open in disgust if given a wave off. Additionally if I were given a cut, I had to take it. The two signals, wave off and cut, were the only mandatory landing signals. Moreover, once engine power was cut, I was not to reapply power.

It was important to add power cautiously in a low-airspeed, low-power, low-altitude situation such as that preceding a wave off. By applying large amounts of power to a 2,000-hp, high torque propeller aircraft, it was virtually certain the engine would overcome aileron control, causing the plane to roll left and plunge into the sea.

Flight deck personnel, affectionately called "deck apes," handled all the night deck activity without a single word exchanged between them. Speed and efficiency was of the essence, as another aircraft was usually close behind and within seconds of recovery.

As soon as the aircraft's forward motion stopped, a deck director signaled me with two raised clenched fists, indicating to me "apply the brakes to both wheels" and keep them on. The director, signaling with two hands alternately pressed together and separated two or three times while keeping the wrists in contact, then directed me to retract the flaps. Simultaneously, a hook runner emerged from the catwalk to disengage the arresting wire from the tail hook. Once the aircraft was clear of the wire, the director signaled me to "retract the tail hook," followed immediately by a "come on" with both hands raised high. He then transferred control to the next director up the line by pointing with both arms in that direction. Once clear of the landing area, the director gave the "fold wings" signal. The plane was then guided to a parking spot.

The barrier, located amidships aft of the parking area, was lowered to allow the airplane to taxi rapidly over it on its way forward. As soon as it had passed, the barrier was raised again.

Invaluable in protecting the aircraft parked in the bow area of the night deck, the barrier was mounted on five foot hinged stanchions. Two heavy cables were held horizontally into position by means of short, smaller cables attached to the heavier cables by clamps and shear pins. If the tail hook failed to catch a wire on landing, the airplane rolled into the barrier, breaking the short cables and allowing the longer cables to pay out and stop the aircraft with minimal damage.

No aircraft was allowed to move on the flight deck except under the control of a director. Hand and arm signals were standard. When the director wanted to move an aircraft straight ahead, he gave the normal come-on signal with both arms. To turn the aircraft, he pointed a clenched fist at the wheel he wanted me to brake and waved forward with the other hand to turn the plane around the locked wheel. The "stop" signal was two raised fists above the shoulders; "shut down" was a finger across the throat. The deck personnel also had a number of arm and hand signals used to manage the operation of the deck.

At night, directors used flashlights with tubes about six inches long that lit up as red rods. The come-on was indicated by moving the red rods rapidly back and forth as with the day signal; to turn, one tube was pointed at the wheel to be braked while the other flashlight waved the plane forward. Two rods crossed overhead indicated "stop." The plane remained stopped until moved forward again with a signal or when the wheels were chocked. The engine shut down signal was given by drawing a red rod across the throat.

In retrospect, it is amazing that the dangerous activities on the flight deck were accomplished without talking, while the sighting of an enemy aircraft often caused bedlam on the radio.

The flight deck was divided into three sections for purpose of deck control. "Fly One" was the launching and catapult area forward under the direction of the flight deck officer and catapult officer. "Fly Two" was amidship and included the island; it was under the taxi signal officer. "Fly Three," the landing area aft, was the province of the landing signal officer and arresting gear officer.

Within these areas and in order to minimize delay or confusion, personnel had identifying colored jerseys and cloth helmets. The former were generally worn over the dungaree or khaki uniform shirt.

Many people crowded the night and hangar decks. Almost all of the aircraft moves were made by handling crews, each comprised of 12 non-rated seaman wearing blue shirts and helmets. The number of crews was dependent upon the size of the carrier. Handling crews were supervised by plane directors who also directed taxing planes about the deck. These petty officers and chiefs wore yellow jerseys and helmets or, in the tropics, yellow helmets and "Skivvies" shirts. Any aircraft not actually moving had to be chocked; the men who handled the wooden chocks and tie-downs wore purple jerseys and helmets.

The arresting gear and catapult crews, normally petty officers, wore green jerseys and helmets. Hook men, whose job it was to disengage the planes arresting hook from the arresting cable, wore the green jerseys and helmets of the arresting gear crew. This job demanded agility and fine timing and required them to sprint across the deck to the aircraft while wearing heavy clothing and padded gloves.

Fueling service crews, or gasoline crews, were responsible for gassing and oiling aircraft both on the flight and hangar decks. They wore only red helmets without jerseys so as not to confuse them with the fire fighters who wore their red jerseys and helmets while stationed around the catwalks and the island, ready for instant action. Their number included the "Hot Papas," shrouded in asbestos suits and helmets, who approached a burning airplane to rescue trapped air crew. Plane captains were squadron personnel and not considered part of the flight deck crew; they didn't wear jerseys or helmets. Brown helmets and jerseys were worn by flight deck sound-powered phone talkers who communicated with the air officer and LSO, linking the flight deck with the ship's control stations.

A medical officer, and two or three hospital corpsman with white helmets or white arm bands with a red cross, covered the flight deck. In addition, a hook observer on LSO's platform signaled whether the hooks and wheels were down. The ordnance men armed the machine guns and loaded the bombs and torpedoes. The hook observer and ordnance men, like the plane captains, did not wear jerseys or helmets.

As is the case with most regulations, compliance varied from ship to ship and with circumstances. Some LSOs never wore jerseys—others slept in them. Photos abound of plane captains and hook spotters wearing jerseys on the flight deck, but such wear was tolerated, not authorized.

This is how it worked aboard all carriers, but for now my mind was on the qualifications ahead. I was selected to go aboard a jeep carrier, the USS Core (CVE-13).

Part of the night carrier qualification requirements was that you were to make three successful day landings followed by two successful night landings. These landings had to be made during the same day. You couldn't make three day landings on one day and the next day make two night landings.

The first night of carrier qualifications was a fiasco. The weather was intolerably nasty. It was only made worse by the shortened deck of a CVE type carrier and with its short draft, it bounced about the Atlantic wind swept waters like a cork. When the junior officers saw the resistance to this operation by the veterans of our squadron, it made it a questionable operation by all the nuggets. It produced a certain amount of fear in our ability to accomplish this assignment successfully.

By day, the skies were gray and cold with intermittent showers. Snow storms and bitterly cold winds prevailed, causing enormous concerns over icing conditions which might hamper operations, not to mention our lives. During the day we all managed to cope landing safely aboard, but when night fell it was a different matter! There was no horizon line available and most of the time it was pouring rain and sleet. The ceiling was under 500 feet and landing lights had been restricted due to the report that many enemy German subs were in the vicinity The only available illumination were small blue lights on the picket destroyers in front and behind the carrier, and the small 12 degree lights on the carrier deck that could be seen only when we were properly in the groove!

I watched the operations as most senior officers carried out their obligations. As a plane would land, another pilot standing in the cat-walk would leap forward and change places with the pilot who had just qualified. I was mystified by the fact that the Corsair would stay in one piece as most landings came slamming into the deck and the plane was under great stress, crinkling airframes and blowing tires. It took an excessive amount of time for just four planes to be landed as wave off after wave off took place. How in the hell were they ever going to get us all qualified at this rate? As many as five passes for some pilots were not unusual. You now could ascertain who the better pilots were and I was surprised to find out the next day that the junior ensigns in some cases did a better job than the veteran lieutenants.

Due to the extreme weather conditions, pilots were given the opportunity to delay their qualifications until the weather got better. However, this would delay our boarding of our future home, Intrepid. Not one officer aboard declined to fly, as we all wanted to get it over with and agreed that it should be carried out as scheduled.

The night I was scheduled to qualify I was standing in the catwalk, waiting for the incoming plane to land—the one I'd use on my turn. To my horror I watched as LTJG Larry Mead approached the ship and then suddenly stalled in the groove, plowing into the fantail with a tremendous crash. He must have died instantly.

Operations were secured for that evening, but the following morning at 0430 I was the first to takeoff in the pitch black, rain-squalled sky, although the weather had gotten worse. If the squadron was going to complete its requirements and meet the time schedule, they had to go ahead with the carrier qualifications.

On takeoff, I tried to discern the horizon line but it was so black and the rain so fierce that it was simply impossible; I had to rely on instruments to guide me. Since I was the first to takeoff I couldn't see any other plane's exhaust flames to guide me. Trying to make my first pass at landing, I crossed directly over the port side of the carrier. Looking down I could see the little lights outlining the deck. I used my clock to time my down-wind leg, make my 90 degree turn to the left and then timing another 90 degree left, parallel to the carrier on my down-wind leg. From there I could see the little blue light on the trailing picket destroyer and make the appropriate turn into the groove, where I could pick up the lights which outlined the carrier deck. From there the LSO in his fluorescent suit of orange and red, lit by black light, came into view, directing me aboard with his glowing paddles.

The carrier was bobbing up and down so much, one moment I was headed directly toward the fantail and then the next I seemed too high. The LSO was holding a "roger' on me most of the time and I had to really believe in him. I thought for sure I was going to duplicate the fatal performance I'd seen the night before. In this freezing weather, sweat was running down my face as I approached the deck. The LSO gave me the cut and I caught the number-two wire. Taking in my tail hook I was sent off immediately and repeated the same procedure for a successful second landing. From start to finish I wasn't in the air over 20 minutes, but it seemed like hours. As I folded my wings and taxied over the barriers, I knew I had qualified!

I followed a pair of lighted red wands and was directed to the starboard side of the ship. The nose of my aircraft was overhanging the edge of the ship on the starboard side when I was given the closed-fist signal to swing my aircraft to the port. I then found myself lined up with the very starboard edge. I was given the cut engine signal and suddenly the director disappeared!

(As Erickson taxied forward, the next Corsair, flown by Fred Meyer, attempted his landing. His plane hit the deck nose first, with his propeller chewing down the wooden deck with sparks flying. The Corsair missed all the arresting wires, and slammed into the barrier in a nose-down vertical position where it teetered and threatened to go over and crush Erickson! Fortunately, the wind over the deck blew the Corsair backwards onto its wheels.)

I thought I was surely going to "buy the farm!" I sat in my aircraft watching in my rearview mirror as Fred and his plane careened down the deck, spitting fire and sparks while the three-bladed prop threw wood chips all over the deck! I had folded the wings of my Corsair and as I looked to my left the exit was blocked. Looking to my right I could only see the black water below. The chance of survival in the cold Atlantic water longer than 15 minutes was nil. The picket destroyers would have a very difficult time finding a downed pilot in enemy sub-infested waters, and the use of spot lights was prohibited. I decided to take my chances and rely on the barrier wires to do their job and, fortunately, along with the wind, they did.

Fred Meyer claimed 22 aircraft had been damaged and several of the junior pilots were severely injured. Later, I found out some of the old fighter pilots from VF-17 had put up a great fight to avoid making the night landings all us junior officers were required to do. Some of our senior officers never did make them!

CHAPTER 32

"Baptism of Fire"

by Roy D. "Eric" Erickson

(Sunday, March 18, 1945)

The rough sea sheared against the great steel hull of Intrepid. I looked up at the black ominous sky and found it next to impossible to discern the horizon from the cockpit of my F4U-ID Corsair.

My plane captain, a man in his early thirties, was near tears and shaking as he came up onto the wing beside me and helped me into my shoulder harness. He seemed much older to me, though, as I inserted the plotting board into its slot and locked it in place. He told me they'd all heard a report that the sky was thick with Jap planes overhead and he feared for his life. I consoled him as best I could, but I had to be about my business.

I flicked on the black light, illuminating my control panel and enabling me to adjust my trim tab settings. From the bridge the loudspeaker bellowed, "Erickson, turn off those God damn lights!" I complied in a flash. How the hell did they know it was me? With the absolute blackout, I couldn't understand how they could identify me in particular. Of course, the men up on the bridge, had every plane and pilot's position carefully plotted on the deck.

The deck of the carrier seemed to explode with smoke, fire, and noise as we all started our planes. Like a blind man reading Braille, I went through the check list, knowing the failure to follow one step could kill me. Tightening my shoulder straps, I followed the lighted wands of the deck officer and moved forward to the left catapult, lowering and locking my wings into place. The deck crew hooked my aircraft to the hydraulic-powered monster. The image of myself as a stone in some giant sling came to mind. As usual, the commander and his wingman were already in the air and LTJG R. H. "Windy" Hill, my section leader, had just been launched off the right catapult. I put my head back against the headrest and raised my arm to show the deck officer I was ready for launch. Lowering my arm, the catapult shot me into space!

Tail End Charlie was my position in the division of four aircraft led by CDR Hyland. ENS Tessier was on the commander's wing and my section leader was Windy. Not only was the CAG's division the first to be launched from Intrepid, it was the first to land and the first to wait as well. We circled above and watched the other divisions launch and join up, grouping in a long waving tail before each division proceeded to their designated target.

All the fighters and fighter-bombers were in the air and now the torpedo planes were taking off from Intrepid's deck. The first TBM took off and sank out of sight below the deck. I watched as he reappeared and saw him land in the water. Fortunately the pilot had managed to ditch safely to one side of the oncoming carrier. The three crewmen scrambled from the still floating Avenger and managed to get into their life boat. No sooner had they gotten their feet out of the drink when the identical thing happened to the next TBM attempting a takeoff. They too had cleared the on coming carrier and were safely getting into their lifeboat. As soon as the carrier and the destroyers had cleared the area, they were picked up by the trailing DD and transferred to Intrepid for the next day's strike. I thought to myself, does this happen every day? If it did, we sure didn't need the enemy to help us. We were doing just fine by ourselves and it wouldn't be long before we would be out of torpedo planes altogether!

Later, I found out that there was too little wind over the deck that day to obtain enough air speed to get the two planes airborne with their heavy loads of torpedoes and fuel. I never again saw or heard of a similar experience taking place.

It was CDR Hyland's responsibility to coordinate all the attacking aircraft within the air group. He was also air coordinator for other great sweeps involving the aircraft from carriers operating with us. One of my duties was to protect his tail.

The sun and mist were breaking over the horizon as we continued to the target over an endless sea. I was nervous with anticipation over what was in store. It was hard to realize that I was actually on my way to attack the home islands of Japan as my eyes kept searching the sky for enemy aircraft and the ocean below for enemy ships or any sign of life. I calculated the force of the wind from the size of the waves and kept track of our course on my plotting board. If we were to encounter enemy planes and I was to get lost in the melee, it would be my only tool to help me find my way home to the fleet.

Flying wing is not like leading the pack. It is an unending juggling of the throttle and working diligently to stay close to your section leader, keeping a constant watch on any movements he might make. The last position in a formation was usually the first one that got picked off from an unknown assailant coming out of the sun, and I was not about to let this happen. I was kept very busy.

My mind started playing tricks on me when I thought I heard strange noises coming from the engine but a quick glance at my instruments told me everything was OK. My imagination kept conjuring up problems that didn't exist. Was I running out of fuel? Were the magnetos firing properly? Were my guns even working? I even practiced grabbing the ring of my parachute just to be sure it was still there!

An hour of monotonous searching and checking had passed when suddenly through the mist appeared our target-Saeki Naval Base, located on the shore line of Kyushu. I forgot all the imaginary problems and concentrated on the target, arming my guns, bombs, and setting my outboard rockets to fire. I checked all the instruments to make sure all was in working order.

My adrenaline was really flowing as we pushed over in our attack. I went to the outside of the formation and a little behind Windy so that I could concentrate on the target. There were parked aircraft lined up on the runway and with the red Jap meatball zeroed in on my gunsight, I blasted away. I was almost mesmerized watching the first plane explode in a violent ball of flame and the second one fly apart as my bullets struck home.

As we cleared the field I saw a tanker cruising in the harbor. Resetting my eight rockets, I fired them in salvo while strafing the tanker. The rockets all smashed into its deck and hull. As I looked around it was blowing and blazing and sailors were diving into the ocean; one less ship in the Jap Navy. We made a few more strafing runs over the airfield and then CAG gave us the thumbs up and turned for home. I had not yet learned to conserve fuel while flying Tail End Charlie, which used much more gas than when flying lead. Pumping the throttle and flying wide in a turn would suck up your fuel all too quickly. The next day, I learned to conserve fuel by slipping under aircraft on a turn and by fine tuning the richness of my fuel mixture until full rich was really needed. However, the four-and-one-half hour flight had nearly depleted all fuel tanks.

Upon returning to the fleet, Intrepid was in the process of launching aircraft, and was unable to land any planes. Many of us were running out of gas. The carrier Enterprise, the "Big E," had just cleared their decks and had turned into the wind and they were prepared to take me aboard. As my fuel gauge showed my tanks to be nearly empty, I knew I would have to make the first approach a good one or go for a swim.

I made the standard approach down the right side of the carrier, lowering my flaps and landing gear and opening my cowl flaps. I lowered the tailhook, making sure my tail wheel wasn't locked, put the prop in full rpm and moved the mixture control to full rich as I made my turn into the groove. Sighting the landing signal officer, I waited for his signals and corrections, but he stood there just as if he was cast in stone. While closing in on the fantail of the Enterprise, I kept watching the LSO. The Enterprise's deck seemed much narrower and shorter than Intrepid, but I had never made such a perfect approach before and I began to wonder if the LSO was OK. He was still standing there with both arms extended in a "roger." As I came abreast of the fantail, he gave me a cut and I dropped onto the deck, grabbing the number two wire. I raised my tailhook, folded my wings, and crossed over the barriers. I was later told that I had five gallons of fuel left. Not enough to have taken a waveoff, but then I knew that already.

Crossing over the deck I went into the pilot's ready room and grabbed a cup of cocoa. As I sipped the hot concoction the LSO arrived, with a big smile on his face. Shaking my hand, he said, "Congratulations, you're the first Corsair I've ever landed!" The Big E was flying F6Fs and this explained his statue-like stance as I made my final approach. He said he figured I knew more about the Corsair than he did, and he decided to leave it all up to me!

(Monday, 19 March, 1945)

Japanese Air Group 343 was flying the new NIK2-J Shiden 21, code name George. Derived from a Kawanishi Rex float plane, the aircraft had gone through many stages of development. Unlike the Zero, the George had a special automatic flap system that enabled it to turn on a dime. It also had self-sealing gas tanks, armor plate to protect the pilot, and two 20mm cannons in each wing. With a top speed of 369 mph at 18,370 feet, it was a formidable foe against the Corsair and the Hellcat. In the hands of a veteran pilot, the George was probably one of the best fighters to come out of the Pacific theater of war.

Dawn was breaking above the mountains, and what a sight it was, nearly overwhelming! Having studied art since the age of eight, I was familiar with Japanese prints. I had always thought Japanese artists had taken a very broad artistic license When they showed their mountain peaks to be so sharply pointed, with clouds of mist and fog lying in milky layers at their base. But here, before my eyes, was the very embodiment of a Hiroshige print, and with the rising sun no less, providing the luscious kind of theatrical lighting. The colors of the canvas before me literally produced tears of appreciation, but then my mind snapped back to reality. I gently rolled off to one side to test my guns. Hitting the charger-buttons with my feet, I pulled the trigger on the control stick and found to my satisfaction that all six fifties were in superb working order.

We crossed the mountains and arrived over the inlet at 12,000 feet. As we approached the target, I could see a group of eight Japanese planes circling 6,000 feet above us at eleven o'clock. Little did I or any of us realize they were from Genda's 343 Air Group, led by LT Kanno from the 301 Squadron.

Excitedly, I radioed Hyland and told him of my sighting, but he informed me he'd already been watching them for the last few minutes. They were tail chasing each other in a circle. One would do a snap roll, followed by another and then another. Whether they were trying to draw us away from the target, getting their courage up, or just plain showing off, I do not know. This was my first encounter with enemy fighters and, staring at the bright red meatballs on their wings and fuselages, it seemed as though I was watching a movie unreel before my eyes. We were indeed over Japan!

I couldn't help thinking of an article that had been recently published in Life magazine. A full page photo showed some downed B-24 pilots getting their heads chopped off in a town square by their Jap captors. The memory alone left me both angry and apprehensive as we flew deeper over the island.

The commander led us across the bay, making a 180-degree turn, and we started our approach toward the oil storage tanks. At about 12, 000 feet, we released our belly tanks and then armed our bombs in the dive. Our division began the attack followed closely by the other six planes. We were met with an absolutely ferocious barrage of antiaircraft fire. Dropping our bombs, firing our rockets, and strafing the target, we pulled out over the bay at 3,000 feet to avoid small arms ground fire, which was as capable of killing us as anything else. Then a peculiar thing happened.

I informed CDR Hyland that I'd sighted a Rufe float plane taking off in front of us. Hyland went after the Rufe and Hill and I started to climb. I was confused—should I join Hyland or continue with Hill? In a split second I realized there really wasn't a decision to make. My duty was to fly wing on Windy, and I stuck right with him.

Hyland splashed that float plane, a well deserved first victory in his new squadron. My observation was that it was a Rufe, a basic Zero fighter with a float attachment. Hyland recorded it as a Rex float plane, the forerunner of the George which we would encounter shortly. Almost 50 years later I was proven right. The CAG's victim had indeed been in a Rufe! LT Shunji Yamada of the 951 Air Group survived the encounter, but his aircraft was destroyed.

My three years of training were about to pay off! No longer did I have to think about a maneuver—it was if my aircraft and I were one! Every fighter pilot aboard ship thought he was the best and knew he was. With a competitive nature and spirit of aggressiveness, he wouldn't allow for the possibility that he would be the one to get shot down. His ego wouldn't allow it. I myself had formed great confidence in two things, my ability to navigate accurately, but most importantly, I could damn well hit whatever the hell I was aiming at.

After we'd climbed to 3,000 feet, I suddenly realized, as had Windy, that we were alone! I kept a very watchful eye on the circling enemy aircraft above us, and as we continued to climb two of the Zekes did a snap roll and flew straight down toward us. As they came within range I pulled up into them, pulling back on the stick, and graying out for a few seconds. Thank God, I had my anti-blackout suit on, for I could still see the oncoming aircraft. Without the suit I would have blacked out completely. My vision was clear as I put my gunsight directly on the lead plane and fired.

The Jap pilots were flying in such a tight section that I raked both planes with .50 caliber rounds. My plane shook as the tracers flowed, and I could see them sparkle against the silver-gray under-bellies of the oncoming Zeros. As the lead plane passed over me he was already in flame, trailing thick black smoke. He was so close I could count the rivets in his wings. Windy was below me and wasn't able to confirm my kill, but now, out of formation, I wisely decided to form on Windy and not follow the plane that I'd lit up.

Making a turn to join up, I saw another Zeke sitting on Windy's tail, guns flashing away! We started our weave and Hill shouted "Shoot the son-of-a-bitch, Eric, shoot the son-of-a-bitch!" By now we were well into our first weave and I answered, "What the hell do you think I'm trying to do!"

My first efforts to get him in my sights were fruitless and I quickly realized that we were weaving too tight. On the second weave, I went out far enough to make damn sure I had him sighted. All this time he had been hammering at Windy and I'm sure my partner's drawers were a bit moist. The Zeke suddenly broke away to the left of Windy and now flew directly in front of me! I could see Windy off to the right, still zigging and zagging, seemingly unaware that his pursuer had turned away. In a matter of seconds I had the unwary Jap pilot perfectly bracketed in my sights and then, with all my six guns blazing, the Zeke blew apart! The front part of his plane flew on straight and level, but the tail section sheared off behind the cockpit and spun crazily away. Gaining on him fast, I flew through all kind of flaming debris, instinctively ducking to avoid getting hit by all the fragments. There was no sign of anyone even trying to jump out of the enemy plane, so I assumed the pilot was dead. The time it took to make that assumption was all the thought I gave it.

I was now high above and in front of Windy and I made a turn to join up on him. To my amazement he was already shooting at another Zero! I watched as his quarry burst into flame and the pilot scrambled out. He was wearing a full length, dark brown flight suit and an astonished expression. As his chute billowed I tried to get my sights on him, but to no avail. At the time I didn't think of shooting the parachute. Later I heard it may not be a good thing to do, as it didn't help the treatment given to our POWs below. I had no moment to consider this either—I was at war.

What neither of us realized was that Windy had just shot down the Japanese ace, LT Kanno, leader of Squadron 301 from Air Group 343!

Joining up, we headed toward the sea. As we traversed the mountains and hills I saw another Zero directly below me and heading in the opposite direction. I thought it would be a great opportunity for an overhead run, but since Windy didn't see him, I again thought it prudent to stick with my section leader.

Cruising back over Shikoku island, I looked over at Windy, who had positioned me a hundred yards to the side of him, and I couldn't believe my eyes. There sat a Tojo on his tail and all four of the Jap's cannons were blinking his way! I shouted a warning to Windy and we immediately broke to weave, but I couldn't get my sights on the enemy with the proper lead. Not wanting to waste ammo, I didn't fire. Having learned my lessons well the first time around I went further out than seemed necessary. Coming back on the second weave, I had a straight 90-degree deflection shot. I had to put my sights directly on Windy's head to get the proper lead on the enemy aircraft, and it took real nerves of steel to pull the trigger. True to my training, the tracers seemed to bend directly into the Tojo! He went ablaze and slid to earth as if he was on a greased wire. In my mind, this was my THIRD victory of the day!

CHAPTER 33

"Sinking of the Yamato"

by Roy D. "Eric" Erickson

It was 1030, 7 April 1945. I was resting in my sack, having served as the duty officer for the early morning 0600 flights, when suddenly, over the squawk box I heard the message, "ENS Erickson report to the ready room!" I put on my pants and shirt and slipped into a pair of loafers. In case I had to go for a swim, I wanted to get out of my shoes fast.

I hurried down the corridor, through several hatches, crossing the hangar deck, and up the ladder to the ready room. The duty officer said, "Get on deck." One of my buddies, ENS Ecker, had injured his hand the day before and was unable to fly, so I took his place. They needed every available pilot. I didn't know who I was flying with and was completely unaware of the urgent situation. Jotting down Point Option on my plotting board, and putting on my flight gear, while noting the deck assignment for the aircraft, I left the ready room.

Pilots were firing up their engines and many were already in the air. I crawled aboard my assigned plane and strapped myself in. The plane captain handed me a chocolate bar and a canteen of water. I said, "What the hell is this for?" Never had I been treated with so much attention. He said, "Haven't you heard? They have located the Jap fleet!" Suddenly, it dawned on me what the huge 1,000 pound bomb was doing under my plane.

I had never seen so much helter-skelter as I was directed forward by the deck officer. He rotated his flag violently and then pointed it down the deck. Pushing the throttle full forward, my plane rose from the deck. I joined up on a division that was missing a plane. I found myself flying wing on LT Wes Hays from Texas. The other two slots were filled by LTJG Hollister and ENS Carlisse. On the way to the target the sky became increasingly black due to rain squalls and the heavy weather front the Japs were using as cover.

At 0830 an Essex search plane sighted the Yamato force steaming south toward Okinawa. The Japanese were then shadowed by a pair of PBM Mariner flying boats which held contact for five hours despite being shot at by their prey. At 0915 Admiral Mitscher sent off 16 fighters to track the Yamato and at 0100 Task Groups 58.1 and 58.3 began launching a 280-plane strike. Included in this group were 98 torpedo-carrying Avengers. The Hancock was 15 minutes late in sending off their 53 plane contribution. We (TG 58.4) followed this main strike with 106 planes.

Still groggy from this unexpected call to duty I cranked off the cap to the canteen and took a swallow of water. I grabbed the candy bar that I had stuffed in a trouser leg of my flight suit and I thought about how thoughtful the plane captain had been. It provided me with a new surge of energy as I slid under the lead plane. Now the rain squalls were getting worse and visibility was lessening. Looking down at the water I could see white caps below and I estimated the wind to be around 25 knots, not a good day to make a water landing. No longer was the engine making imaginary noises as on my first combat flight, but was purring like a kitten. Checking all the instruments the plane seemed to be functioning properly. We had traveled for over two hours searching the sea through this muck looking for the elusive Japanese Fleet. I hadn't been present at the briefing so I had no way of knowing exactly where we were headed, but by plotting the time and course I knew we were somewhere south and west of Kyushu.

At about 1330 my skipper, LCDR Rawie ("Red One"), was

about ready to turn back. We were flying at 1,500 feet when, suddenly, through the scud, directly beneath me I saw a gray massive structure. I was the first in our group to see the biggest damn battleship in the world—the mammoth 64,000 ton Yamato! It had been hiding under rain squalls and low clouds. I transmitted the message to "Red One" that the Yamato was directly below and Wes Hays signaled us to start our attack! We whipped into a fast 180-degree turn in an attempt to get on the Yamato! As we broke through the 1,500 foot ceiling, the Yamato appeared to be almost dead in the water, but still in a slow left turn. Smoking destroyers were all over the place and only two could be seen swiftly maneuvering through the water. It was a Navy pilot's dream with no enemy aircraft to repel our attacks.

I had watched our task force shoot down Kamikazes like they were ducks in a shooting gallery and I thought, "Oh my God! I'm now the sitting duck!" Now, I know how a Kamikaze pilot must have felt as he was preparing to make his final assault. How could all those ships down there miss when they were armed with all that sophisticated radar? It was a true test of courage! Even the Yamato's 18.1 guns were shooting at the approaching aircraft (as they had in vain at the flying boats). In addition to her big guns, the Yamato was able to fire on us with her 24 five-inch guns and about 150 25mm guns. The light cruisers and the destroyers joined in on the crescendo!

We tried to get our sights on the battleship, but we had started our run so low it was impossible. I could see men scrambling all over the deck at what looked like mass hysteria! Where were they all going? Diving and pouring on the juice we crossed over the Yamato and strafed the hell out of it. I could see bodies flying all over the place! In return the sky was bursting with thousands of brass wires as the Japs' guns zeroed in on us! Looking down I wondered why I wasn't getting hit, the tracers were so close you could smell the cordite! Black flak bursts were bouncing my plane violently from side to side and the sky was turning dark! I thought for sure this was the day for me to meet my maker!

I could read the wake of the light cruiser Yahagia, an Oyodo class cruiser, as it turned around toward the Yamato to help protect it. This was my first time to wing Wes, but I knew he was heading directly toward the cruiser! I moved in closer and closer on him and concentrated on his aircraft as we dropped our bombs in unison! After releasing our ordnance, we headed for cloud cover and then, as if on a roller coaster, we dove back down and skimmed along the ocean floor strafing the destroyer Isokaze that lay dead ahead. Flashes of bright light were blinding us as the destroyer tried to elude our attack. Suddenly the destroyer stopped firing as it went ablaze and dark black smoke poured from its deck. We passed over it and pulled up again into the low cloud cover. We thought we were out of range of enemy fire and as we looked back, no longer were the cruiser and destroyer in view. We circled at 5,000 feet and five miles from the Yamato. The clouds started to clear and we could see the battleship and the rest of our group making their attack.

While we were circling, I noticed great spouts of water rise from the ocean floor. My first thought was, some of us hadn't dropped our ordnance, and were now doing so, but this was not the case. The damn Yamato was still shooting their big 18.1 guns at us—the largest guns in the world! Then, Air Group Ten's dive-bombers, torpedo planes, and fighter-bomber pilots completed their run and a terrific explosion took place. Great billows of black smoke were sent skyward over 6,000 feet—the end of the biggest battleship in the world!

The giant warship listed heavily to port and at 1423 disappeared underwater, followed by explosions of rupturing compartments and her magazines. It had taken ten torpedoes and five direct bomb hits to sink the Yamato. Her sister ship Masashi had required 11 torpedoes and 16 bombs to send her down in the Sibuyan Sea the previous October.

Two light cruisers in the Yamato force were sent to the ocean floor. One destroyer was sunk outright; three others were so severely damaged that they were scuttled. The four other destroyers were damaged to varying degrees. Only 269 men survived the Yamato; 2,498 including her captain and the force commander went down with her. Almost 1,200 more men were lost floundering in the sea. With the light cruiser and destroyers that were sunk, 3,700 lives were lost in the greatest Kamikaze sortie of all.

In an incredulous comparison, TF 58's carriers lost three fighters, four dive-bombers, three torpedo planes, and a total of 12 fliers. One of the aircraft, a Corsair, was lost in a midair collision en route to the attack.

Our air group rendezvoused all its planes and headed for home. Not one of our aircraft was shot down, and only a few tail feathers were lost in this auspicious attack.

Back in the ready room aboard Intrepid, five hours and fifty minutes later, our division was asked to identify who hit and who missed the cruiser we sank. The pictures taken by the photo planes showed three hits and one near miss. Of course, trying to identify your bomb from the others would be impossible. The four of us dropped our bombs together at the lead of our division leader and then, immediately pulled up through the low lying clouds. One of the pilots however, said, "I saw my bomb hit!" The remaining three of us were asked to draw cards to see who had dropped the bomb that nearly missed the cruiser; low card would receive that honor!

The three of us walked over to the gaming table still set up in the pilots' ready room. We asked one of the pilots who was mulling around to shuffle the cards. He placed the cards on the table and asked if we wished to cut. One of us cut the cards and he replaced the cut under the pack. I grabbed a section of cards as did the others. Turning the cards over, I discovered I had drawn— THE DEUCE OF HEARTS!

At the time I did not realize its importance and I thought, "The whole damn war is like this, it's the luck of the draw!" A near miss in my mind might have done more damage then a direct hit. A bomb at the water line could have blown open the seams, and may have been the bomb that sank the ship!

No one told us the low card would receive the Distinguished Flying Cross, and the high cards would receive the Navy Cross! The Navy Cross was one of the most coveted awards the Navy could bestow. A pilot can spend his entire career flying for the Navy and never ever have the opportunity to receive such a distinction. The consequences of this draw continues to gnaw at me to this very day!

AIRCRAFT CARRIER LSO SIGNALS

Drawn by Roy D. "Eric" Erickson
(Used with permission)

PART VI

Wallace Bruce Thomson

I was born in the small town of Hasbrouck Heights, NJ Sept. 6, 1917. That was my residence until 1939. My father was the supervising principal of the three schools in that town and was a prolific writer of school text books. Adjacent to our town was Teterboro Airport which was a center for aviation pioneering in both mechanical innovations and global explorations in the twenties and thirties. Charles Lindbergh, Admiral Richard E. Byrd, Clarence Chamberlain, Clyde Pangborn, Bernt Balchen, Anthony Fokker. Most of these men were later inducted into the New Jersey Aviation Hall of Fame. Later on, in 1979, my own brother, Johnny Thomson, after an impressive aviation career, was inducted into that organization.

In 1939 I graduated from Montclair State Teachers College, majoring in Physics and Chemistry. For the next two years I taught science at the Jamesburg, NJ high shool. In 1940 and 1941 I earned my pilot's license, taking the primary and secondary Civilian Pilot Training courses in Aeronca and Waco planes.

In Sept. 1941, I entered the Navy and began flight training at Floyd Bennett Field at Brooklyn, NY. I became an Aviation Cadet at Jacksonville, FL Naval Air Station in Jan. 1942. In Aug. 1942 I was commissioned as Second Lt. in the Marine Corps at Opa Locka, FL, having completed training as a fighter pilot.

In Sept. 1942 I was shipped to Ewa Marine Corps Air Station on Oahu, Hawaii and began training in Grumman F4F Wildcats in Marine Fighter Squadron VMF-221, which had been decimated at Midway the previous June. In Nov. 1942 I was sent to the

Palmyra Island Naval Base to defend that base, flying the F4F Wildcats in squadron VMF-211, which had been wiped out at Wake Island the previous December.

In June 1943 we returned to Hawaii and began training in F4U-1A Corsairs. In Aug. 1943 we shipped out for the Marine Base at Turtle Bay, Espiritu Santo in the New Hebrides. In Oct. 1943 we arrived at Turtle Bay and a few weeks later we were sent to the Russell Islands in the Solomons.

On Nov. 1, 1943 we provided air cover for the invasion of Bougainville by the Third Marine Division, fighting off repeated attacks by Jap planes based at Rabaul. By late Dec. 1943 we were based at the Torokina airstrip on Bougainville and began making escort and fighter sweep attacks at Rabaul on New Britain Island. On Jan. 3, 1944 I accompanied "Pappy" Boyington on a fighter sweep to Rabaul during which he was shot down and imprisoned by the Japanese. During this month I shot down two Japanese Zeros and damaged a third.

In Feb. 1944 we were based on Green Island halfway between Bougainville and New Ireland, where our flights to Rabaul were shortened. In June 1944, after 21 months of overseas duty and 61 combat missions, I was returned to the States and sent to Marine Corps Air Station, Cherry Point, NC. There, in VMF-911, we formed the first day-fighter squadron flying the new Grumman F7F Tigercats. In Feb. 1945 I was sent to Naval Air Station Patuxent River, MD to be a test pilot, flying the F4U-4s and the F7Fs. In May 1945 I rejoined my squadron in NC and was soon sent with them to MCAS El Centro, CA. In Aug. 1945 we were about to be sent overseas in the F7Fs when the war ended.

For the last three years of the war I was Engineering Officer of my various squadrons. During much of my life I have been called "Wally" but many people also refer to me as "Tommy." I was discharged from the Marine Corps in Dec.1945, having the rank of Captain. While I had many what you might call "narrow escapes," I never wrecked a plane. I did suffer a broken arm in flight and had two bouts with malaria.

In Oct. 1947, after graduate studies in Physics and Mathematics, I joined the Grumman Aircraft Co.as an Aeronautical Engineer. A year later I joined Fairchild Aircraft and Engine Co.in Oak Ridge, TN as a Physicist in the Nuclear Aircraft Program. In 1951 I married Jane Beale, a chemist from Texas. We had three children.

Later, with General Electric in Cincinnati, I became head of Advanced Design in the Nuclear Aircraft Program. In Aug. 1961 I joined the Martin-Marietta Corp. as Manager-Conceptual Design in their Corporate Offices.

In Aug. 1963 I went to the North American Aviation Corp. (later Rockwell and still later Boeing) in the Los Angeles area to perform research on nuclear space power, the Apollo Moon Program, and solar energy programs. I retired from Rockwell International in Mar. 1983 and at that time was selected as the Rockwell Engineer-of-the-Year for 1983. By that time I had six patents on such devices as: nuclear reactors, gas turbines and solar energy storage systems.

I have lived in Northridge, CA since 1963. My first wife, Jane, passed away in 1989. I am now married to Myrtle. Myrtle and I have known one another from first grade on through high school and were friends in the twenties and thirties.

Since retirement I have kept busy doing research on gas turbines and solar energy devices, and keeping the handicaps of my golf group on a computer. We have done extensive traveling. I have given a number of speeches on historical subjects.

CHAPTER 34

"The Russell Islands"

by Wallace B.Thomson

Thirty-eight of us were transferred to a combat area in the Solomon Islands. We were attached to Marine Air Group MAG-21 and later (Nov. 17, 1943) MAG-24. This was to the North Strip on the Russell Islands. We were impressed by the beautiful coral runway built by the Seabees through the large coconut palm plantation reputedly owned by the Lever Brothers.

The Russell Islands were west of Guadalcanal, about 50 miles from Henderson Field and part way up the Solomons slot in the direction of New Georgia. The invasion by American forces was started Feb. 21., 1943 and was conducted by the 3rd Marine Raider Battalion and part of the Army's 43rd Division.

The Military Situation in the Solomon Islands

Even before the start of hostilities in the Pacific, the Japanese war planners had aimed to seize the Solomon Islands and eastern New Guinea. They recognized that the excellent harbor at the north end of New Britain, namely Rabaul, was essential to carrying out their strategic plan.

As part of their astonishing sweep into the South Pacific and Southeast Asia, the Japanese occupied Rabaul on January 23, 1942. Rabaul, on the northern tip of the large island of New Britain, had a fine, deep-water harbor (Simpson Harbor) in an ideal position to cut the lines of communication between the United States and its important ally, Australia. Since July 1942, Roosevelt, Admiral King, General MacArthur, Admiral Nimitz, and Admiral "Bull" Halsey argued about how we would handle Rabaul. At first it was decided that we would invade Rabaul, which would have been a major and costly campaign. About a year later, it was decided to bypass Rabaul, but at the same time, destroy its effectiveness by means of carrier and land-based air attacks.

To get at Rabaul with our full and developing air power, it was deemed necessary to develop air bases on Bougainville so that fighter planes, dive bombers, and torpedo bombers could be within effective range of that harbor. Bases on Bougainville could be from 150 to 250 miles from their target, motivating an invasion of that island.

Our First Combat Flights

Our first flights in the Solomons combat area were chiefly escorting bombers and transport planes in the Central Solomons, that is from the Russells NW to southern Bougainville. On Oct. 22, Thornton and Major Murto shot at a couple of Zeros and may have nicked them. Things were starting to heat up.

It is ironic that the first casualty in our group was caused by a vehicle involving one of our pilots, Lt. Joe Lytle. A description of Joe's injury would make anyone wince. A vehicle carrying our pilots collided with an oil truck from another unit. There was a large lacerated wound in the left thigh exposing the femoral artery and extending from the scrotum to the anterior iliac spine. Joe was in shock and was treated with plasma and morphine. Nasty wounds are a dime-a-dozen in combat but it is something to see it up close. Joe was shipped to a hospital at a rear base—we never saw him again.

We Cover the Bougainville Invasion

For the last few days in October we were told that there would be an invasion of Bougainville Island and that we would supply air cover. From the Russells, Bougainville was beyond our effective range, so we would have to stage through Munda or Vella Lavella refueling at those airstrips.

On the morning of Nov. 1, 1943, sixteen of us took off for Vella (the airstrip was usually called Barakoma). I was in the last division, my wingman being Lt. McCaleb. Normally we would try to use up the gas in the internal wingtip tanks, then purge these two tanks with carbon dioxide so that in combat they would not explode. The gas in these tanks would last roughly a half-hour then we would switch over to the main 230-gallon tank between the pilot and the engine.

I had just emptied and purged my wingtip tanks when out of the corner of my eye I saw McCaleb abruptly drop out of formation. My first thought was that he had forgotten to switch tanks. But his engine didn't start and it looked like he would ditch in the water near a small island. I pulled up to the division leader, Major Buck Ireland, and attempted to signal that I was going back to cover McCaleb. Buck nodded, apparently not seeing what had happened to Mac.

By the time I caught up with McCaleb he had made a water landing, was floating in his Mae West and was fumbling with his life raft. I slowly circled and he waved to show that he was OK. I continued to circle wondering what to do when a PT boat or crash boat appeared in the distance from a direction I judged to be Munda or Ondonga. If McCaleb radioed for help I sure didn't hear him. I checked out the U.S. flag on the boat and could see that it was heading straight for McCaleb. At that point I headed for Barakoma hoping to rejoin the rest of the guys.

Landing at Barakoma I was able to join them, and gas up for our mission to Bougainville. I was startled to realize that the rest of our formation didn't see McCaleb go down, didn't see me go back to help, and didn't seem worried that both of us were missing.

Our chief mission for that day was still ahead of us. The Third Marine Division was invading Bougainville at Empress Augusta Bay close to a place called Cape Torokina. There were supposed to be about 40,000 Japs on the island. There were several Jap airstrips at each end of Bougainville plus the five airfields at the fortress of Rabaul. For all we knew, the Japs might attack our invading forces with planes coming from all directions.

The flight from Barakoma up to Torokina took about 45 minutes. We had to circle out to the left then back to the right to avoid the five or six Jap fields at places like Ballale, Shortland, Buin, Kahili and Kara. About 12 of us finally arrived on station. Far below us we could see dozens of landing craft making curving white wakes as they sped toward the shore. From our vantage point up there at 20,000 feet or so things seemed to be going smoothly. In reality (as we later found out) the Marines were having a terrible time. Landing craft landed at the wrong places. Vital supplies were lost, the Japs raked the invaders with 75 mm artillery fire, the shore was open to rough seas, and troops found themselves waist deep in muck.

The good news was that the Japs were surprised, not so much that we had invaded, but that we had chosen Torokina—no harbor, no good place for an airfield, and all those dismal swamps.

Ward Hower, my usual wingman, was back on my wing. We climbed to 25,000 feet to look things over. In a situation like that you don't know whether to fly high or low. If you fly low to protect the troops, the Zeros get a vital altitude advantage. If you fly high, bombers and strafers could get at the landing craft and troops and you never know it.

We circled the invasion area for an hour. No Jap planes showed up. We could only guess what was going on down on the ground. We kept circling, surprised that we didn't see any planes at all, friend or foe. Where were all the dozens of Corsairs we had seen at Barakoma?

Finally I saw about a dozen "P-40's" a few thousand feet above us. My first thought was, "Boy, those P-40's sure are flying high, must have those new Rolls-Royce engines in them." They also were all strung out in an unfamiliar formation, maybe they were New Zealanders.

Then the leader waggled his wings and started to dive at us. I went from 1700 RPM to 2600 and to full throttle just like that. Then went into a 45-degree dive. Ward followed suit. Suddenly Zeros were all around us. Their big red meat balls flashed angrily in the sun. If they fired, I didn't see any tracers. We knew Zeros couldn't dive with the Corsairs, especially if they feared that other American planes were down there.

Their attack ended as quickly as it had started. The Zeros disappeared for good. It was our first contact with the enemy and it was an ignominious retreat for us. But at least we were OK. We finished our patrol at 15,000 feet, nothing happened. Then we returned to Barakoma, gassed up and flew to our base in the Russells. Gulping all that oxygen really made me hungry. I ate enough for three people.

As I write this I am looking at our squadron "intelligence" record for that day. It says that McCaleb had "engine trouble" and furthermore "none of our pilots had any contact with the enemy". I would guess that McCaleb simply forgot to switch from his wing tanks back to the main tank. Also, Ward and I being jumped by a dozen Zeros certainly would qualify as "contact with the enemy". It is really surprising that there were no more enemy contacts as records show that 104 Jap aircraft attacks were made on the Torokina beachhead during the first 24 hours of that invasion.

CHAPTER 35

"Several Crashes"

by Wally Thomson

I Have Trouble Starting an Engine

A day or two later a bunch of us were assigned to patrol Munda airstrip. At that point in the Solomons campaign the Japs were only a half hour flight time away from Munda, as they were based in the southern Bougainville area. My Corsair refused to start. Again and again I fired the shotgun starter, the prop would turn over a few times, but the engine wouldn't even cough a little.

I decided that there was something in the starting procedure that the engine just didn't like. As a desperate measure I decided to reverse every cockpit setting for the usual starting to see if the engine would show at least some sign of starting. I put the mixture control in manual lean, the prop in high pitch , the booster pump was turned off, etc. And lo and behold the engine started right up. Then I put all these controls back in their proper positions and warmed the engine up thoroughly. The RPM drop on the right and left mags (magnetos) was normal and I could rev the engine up to 2200 RPM.

I was satisfied that the engine was running OK and took off for the Munda patrol with the other guys. The engine did operate perfectly and brought me back to the Russells North Strip without incident.

The reason I am going through all this was shown on the following day. That day was my day off but I was down at the flight line watching the planes being started for the day's mission. Our pilots were to do a patrol over southern Bougainville. Looking down the long line of planes, I could see that one near the far end would not start. It was the same one that I had trouble with the previous day.

I decided to walk down there and see if I could help them. The pilot was my old friend Lt. Bob Hatfield. I yelled at the mechanics over the roar of the other planes that I might be able to start it. They looked dubious but gave me a smile and helped me into the cockpit. I promptly reversed everything just like I had the day before. To everyone's surprise, including my own, the plane started up immediately. I turned the ship over to Bob Hatfield and he went off on his mission to Bougainville.

And now for the bad news. They ran into a hornet's nest of Zeros up the slot and poor Hatfield was shot down. All my so-called help had only led to disaster.

The "Barrage" of November 8, 1943

Again it was my day off. Why should bad things happen on your day off? We pilots had a little ready hut made of coconut logs sitting in the coconut grove a short distance from the North Strip runway. Ward Hower and I were outside playing chess as usual. There were two or three other pilots lolling around. As usual the air was balmy and humid but with no threat of rain. The scene was Hollywood idyllic.

Suddenly there was a sharp crackling sound like Chinese firecrackers above our heads. We jumped but then sort of froze in place. Then we saw palm trees topple one by one, chopped in half at about mid-height. At the same time a column of black smoke billowed up from across the runway.

Somebody (it was probably me) yelled, "They're shooting at us, duck!" Then we all hit the deck. Well, the "deck" I hit was a warm and gooey mud puddle. It was like having a nightmare about diaper changing but it felt wonderful just getting down that low to mother earth. The other guys were flat on their tummies. The "shooting" went on for what seemed like 20 minutes, dwindling down to an occasional "pop!"

Finally, we stood up and surveyed the situation. The column of smoke was now just a wisp. I walked back to our quarters, a half-mile away, to clean up and try to find out what had happened. Soon the story was all over. It wasn't the Japs after all. It seems that a crew chief was in the cockpit of a Corsair when a gasoline fire at or near the engine burst out enveloping the cockpit.

In his desperate efforts to get out the poor guy pushed the throttle all the way forward which blasted the flames away. At the same time he made a frantic jump out of the cockpit. I believe he escaped injury but the plane roared off across the taxi-way and crashed into the palm grove. The fire enveloped the whole plane. Soon, from the heat or for some other reason, the .50 caliber Brownings began to fire, strafing the area. The plane was aimed right at our ready hut, but just above our heads. Its distance from us was about 250 yards.

The group "intelligence" report on this was a terse, "An F4U, Bureau No. 17730, exploded and burned on the ground while being started by the plane captain on a routine check. Negligence was not apparent".

Several Crashes During November 1943

Munda Point was at the western extremity of the New Georgia Group. The Seabees built a coral airstrip there in mid '43. My own experience with Ondonga was that it was a little small, was unusually congested, mostly with Corsairs, and had a large number of potholes in the runway and taxiways.

My good friend, Dave Escher, put a Corsair in the drink just off the end of Ondonga. He was only a little shaken up. Then on Nov. 11, Lt. Fidler crashed on the Ondonga field. He also was not injured. Still later, on Nov. 16, a Lieutenant, who shall remain unnamed, also crashed at Ondonga. He was reportedly uninjured. For some reason, however, he was sent to a base hospital for about three weeks. Whether this was for physical or psychological reasons, I do not know. Later he was transferred to another unit.

A more spectacular crash and one to which I was witness, occurred at the coral North Strip at the Russells on Nov. 16. Lt. McCaleb came up to me and asked if he could borrow my parachute and back-pack (emergency supplies, such as a mirror, water, morphine, mosquito netting, etc.) for a short flight he was scheduled to make. I was happy to comply but wondered about his unusual request. Then I recalled that he had put a plane in the drink west of Ondonga two weeks earlier. Maybe he didn't have his replacement gear yet.

So Mac took off on his short flight which was officially called a test hop. About an hour later Ward Hower and I were standing near the runway watching Corsairs takeoff and land. Suddenly I saw a Corsair with number 826 on it come in over the runway in what looked like a normal approach. Seeing the 826 number, I turned to Ward and said, "There's McCaleb. He's wearing my back-pack. I hope I get it back." Just as I said that McCaleb's Corsair disappeared momentarily behind one of the buildings alongside the runway.

When his plane reappeared from behind the building, we were astonished to see it was upside down, traveling tail-first, and had not yet landed. The plane landed in that horrendous position and, gouging a deep furrow in the coral, slid down the airstrip for what seemed like a half a mile. McCaleb, fortunately, was not seriously injured. He was hanging upside down, plastered with coral and unable to immediately free himself.

The nearest people to him were a bunch of Seabees. He was elated when several of them reached into the half-buried cockpit and tried to get him out—or so he thought. Then he was flabbergasted when, instead of rescuing him, they tried to take the watch off his wrist! It seems they thought he was dead and that they might as well get a little loot. Of course, when Mac let out a muffled yelp, they immediately started to drag him out of the cockpit.

When they managed to extricate him they apologized profusely about the watch. McCaleb took it in stride and all was forgiven. The plane, however, was considered a total loss. Thus in two weeks McCaleb had lost two Corsairs. He was well on his way to becoming a "Japanese Ace". I never heard that he lost any more. But in my career I knew two Corsair pilots who could claim to be "Japanese Aces", having each crashed five Corsairs.

CHAPTER 36

"Boyington's Last Flight"

by Wallace B.Thomson

Bougainville is the most northwestern large island of the thousand-mile long Solomon Island chain. It is about 125 miles in length and averages about 30 miles in width. Down its center runs a spine of very rugged mountains including the smoking, active volcano, Mt. Bagana, at some 10,000 feet in elevation.

The heaving ridges and gullies, the dense tangle of jungle vines, the oppressive tropical humidity, the swarms of insects, the torrential tropical rainstorms, the waist-deep swamps, and the threat of tropical fevers, all made Bougainville a wretched place to conduct military operations.

But on Nov. 1, 1943 the Third Marine Division under the command of Marine Major General Turnage invaded the southwest coast of Bougainville at Empress Augusta Bay. The reasons for choosing this location were to construct air strips close enough (200 miles) to Rabaul so that fighter planes could manage to fly from this bay over to Rabaul and return, to avoid Japanese military forces concentrated at the south and north ends of the island, and to locate where the construction of airstrips and radar sites was a feasible project for the Sea Bees.

The Third Division disembarking from a dozen transports assaulted the beaches in the Empress Augusta Bay at Cape Torokina and Puruata Island. The surf ran high, landing craft landed at the wrong beaches, boats broached in the surf, and Japanese shellfire killed many Marines. By nightfall several beachheads were secured, although many of the troops ended up waist-deep in swamps all night.

I myself had patrolled the invasion on D-day, Nov. 1, 1943. My wingman, Ward Hower, and I were attacked by about a dozen Zeros that afternoon but we managed to dive away. From above, the landing craft were a stirring sight. It was only much later on that we learned that the Marine ground troops were taking heavy shellfire and had suffered many casualties.

Within a month the SeaBees had constructed an airstrip at Cape Torokina complete with a steel-lattice Marsten Mat runway. By late December we were based at this new Torokina airstrip living in four-man tents in a forest of huge banyan trees.

It was New Year's Day 1944. Two days earlier our first flight over Rabaul, 200 miles to the west from Torokina, followed by a Dumbo (rescue) flight looking for a downed B-24 Liberator, promised a grim and difficult month of January. On New Year's Day things were easier for me as I was assigned to a routine patrol flight over the Torokina beachhead. It was routine in that no Japanese fighters or bombers showed up.

On the morning of January 3rd, 1944, we were awakened in our tents in the jungle at 5:00 A.M. Our mission that day was to take part in a 48 plane fighter sweep across the Solomon Sea to the vast fortress of Rabaul. We pulled on our flight suits in the semi-darkness. Then we banged our boots on the tent floor to knock out any poisonous centipedes that might have crawled in there during the night. (I never saw any but some guys were bitten.) No shaving, no showers, no brushing of teeth.

We shuffled through the giant, towering banyan trees to the makeshift mess hall 50 yards away. My breakfast consisted of two large cups of coffee and one slice of something that resembled French toast. That was my last meal till evening.

Outside on the dank and muddy road some olive-colored vans that were called Carry-Alls were waiting. They bumped all of us on these newly bull-dozed paths down to the large command post by the Torokina airstrip. The post was built half underground and had coconut logs for side walls. Inside eighty or ninety officers were gathered in the gloom—fighter pilots and Marine Air Group Headquarters brass.

Two or three colonels gave little talks on the upcoming mission. In the dim light they and their maps were hard to see and their voices hard to hear. Then a chubby, yellow-skinned officer arose to complete the briefing. There was a tense silence immediately in the post. The man looked overweight, drawn, humorless, almost suffering. It was Pappy Boyington—to those close to him it was "Gramps". The famous Boyington, who had 25 Japanese planes to his credit—two behind Joe Foss. Boyington, who had just four days left before being shipped back to the states. Boyington, whose VMF-214 "Black Sheep" squadron had lost six pilots in the past ten days. Boyington, who had shot down four Japanese planes in the past week in a desperate effort to catch Joe Foss. Boyington, who had to put tobacco grains in his eyes to keep awake, and who was covered with a gross tropical skin disease that we all called "the crud".

Pappy was describing our mission. It was a fighter sweep to Rabaul. Rabaul, the deep water harbor at the north end of the large island of New Britain. Rabaul had been taken over by the Japanese from a small Australian force Jan. 23, 1942 and built into a major naval base with five airfields surrounding it. General MacArthur had strongly advocated taking Rabaul by force. But, probably as a result of Admiral Nimitz's insistence, the strategy of by-passing Japanese strongholds like Rabaul and Truk was adopted. Rabaul, it was decided, would be neutralized by air power—Army, Navy, Marine Corps, Australian and New Zealand. By Jan. 1, 1944 Rabaul was near the peak of its considerable defensive strength. Zeros swarmed around the place from its five airfields.

Just four days earlier I had made my first flight over Rabaul. We were escorting Army Air Corps B-25s that had taken off back on Guadalcanal some 600 miles SE of Rabaul. We wove tightly over the B-25s and kept the Zeros away—but we saw a lot of them out there. The estimate was about 60 Zeros that day.

Today it would be a 48-plane fighter sweep. Four to twelve planes from each of several Marine Corps squadrons: VMFs-211, 212, 214, 216 and possibly one or two others. I had friends from flight school in all those squadrons. Our group had three four-plane divisions that were led by Major Ireland, Captain Langenfeld and Major Hopkins. I led a section behind Herb Langenfeld. On my wing, as usual, was my tent-mate, Ward Hower. Flying with Hopkins was my other tent-mate, Adolph Vetter—soon to make Captain, soon to get his first Zero, and soon to die. The mission was a rather simple one. We would fly in toward Rabaul stacked up between 20,000 and 30,000 feet. I was assigned to be in the highest group. The Zeros might rise to attack us or simply stay on the ground. With their good high altitude performance they might be up in the stratosphere just waiting for us.

Pappy finished his pilot briefing in about ten minutes. To me he seemed to have physically deteriorated in the past couple of months. Then we all broke up and hurried in the early dawn to our long lines of F4U Corsairs strung out along the Torokina strip, wet from last night's rain and streaked with mud. Behind us was the tumultuous Bougainville jungle, where even now we could hear distant gunfire between the Third Marine Division and the Japanese forces surrounding our little salient. In front of us the gray waters of Empress Augusta Bay. To our right off the western end of the runway was the little island of Purapata.

With cartridge starters muffled sounds, the blue and gray Corsairs started one by one. Engines were warmed up, magnetos were checked, wing-flaps tested, cowl-flaps cracked open slightly, seat height adjusted, mixture control set at automatic rich, prop in full low pitch, seat belts snugged up, oxygen mask in place.

Then one by one the Corsairs roared down the Marsten-Mat-covered Torokina airstrip from east to west. They made a giant, climbing circle to the left over the Bay. As we were providing high cover we were the last planes to take off. We cut across the circle and gradually closed in on Boyington and the leading planes. Finally, in a compact group, we headed for New Britain and the fortress of Rabaul, some 200 miles to the west.

Most of the time it was business as usual, but occasionally I would look around at the beautiful yet ominous planes around me and wonder, "What in God's name are we doing here? Why are we here at the ends of the earth in a death struggle with an enemy that we have never known—an enemy just as lost as we are in this tropical hell?" Then the smell of the rubber oxygen mask, the maze of gauges in front of me, the vast stretches of sea and clouds and sky would bring me back to reality.

At first our course took us well away from mountainous Bougainville. For all we knew, Japanese coast watchers were watching for our missions and could quickly radio on to Rabaul vital information. Then we roughly paralleled the southwest Bougainville coast as we slowly climbed. At about 10,000 feet we augmented the main supercharger with the second supercharger called "Low Blower" which we turned on to push more oxygen into our Pratt & Whitney engines.

Soon we climbed through a cloud layer, then another, then still another. It became difficult for those of us performing high cover to follow those below. At 18,000 feet or so we turned the supercharger to "High Blower" to feed the maximum amount of air to our eighteen cylinder engines.

At about that time I swung out a little way from the others and fired my six 0.50 caliber Browning machine guns, giving them a short burst. Some pilots did not take that seemingly essential precaution. The familiar "chug-chug-chug" of the guns and the curving tracers were reassuring.

We passed the island of Buka at the northwest end of Bougainville where the Japanese had several airstrips. We couldn't see them because of the several cloud layers. Just ahead of me was Capt. Herb Langenfeld and Lt. Watson flying a two-plane section. About 50 yards off on my wing was Ward Hower, my faithful wingman with whom I had been flying for the past 14 months.

Occasionally we would see some of the other Corsairs below us but the multiple cloud layers made visual contact difficult. Now we were crossing the Solomon Sea aiming for the northern perimeter of Rabaul. We didn't think of it at the time but to the left of us was the yawning New Britain Trench where the ocean reached a depth of 30,000 feet.

The multiple cloud layers continued to plague us. Boyington and about half of the 48 planes had long since disappeared. Hopkins' four-plane division also was no longer to be seen. We must have been well across the Solomon Sea but there was no sign of Rabaul on our left nor New Ireland on the right.

Soon, I heard what must have been Pappy Boyington's voice. "Let's go down, let's go down", he said over and over in a calm and measured tone. So the eight planes I was with nosed down through one cloud layer but still were well above 20,000 feet. The radio suddenly became full of some sort of frantic, staccato-like yelling that was quite unintelligible to me. Much of it was probably in Japanese. We milled around between various layers of clouds but never went below 12,000 feet.

Finally, we broke through the lowest layer and there was only water below—we hadn't quite reached Rabaul. But there was still action going on as the screaming and yelling over the radio continued. Now I could only see only our own four-plane division headed by Capt. Langenfeld. Looking around for the next half hour we saw no more planes, neither Japanese nor ours.

It would be nice to say that we found our buddies and rescued them from the Zeros but that's not the way it happened. As I learned in my two years in the Pacific, some missions just get fouled up. Of our twelve guys, Major Ireland and Capt. Hopkins each shot down one Zero but the rest of our fellows had no contact with the enemy at all.

Finally, we straggled back to Torokina, not feeling too badly because we all had seen this sort of thing before. The weather and poor communication between squadrons had blunted many a mission. One by one we dropped down to 1000 feet heading east across Empress Augusta Bay, lowered our wheels, circled slowly to the left, received a green light from the newly-built tower, and landed on our little Torokina strip.

But an hour later, when we heard that Pappy Boyington and his wingman, George Ashmun, had failed to return, it became more than a mission foul-up. We had lost a superb leader in Boyington and a real nice guy in Ashmun. A real shock wave went through the pilot's camp in the jungle above Torokina.

Most of us assumed that Pappy and George were dead, although the possibility that they were prisoners was also thought about. The bulk of the 214 squadron—the Black Sheep—was due to be shipped back to the States in one week. Soon afterwards, fourteen pilots from VMF-214 who had not yet completed their tour of duty were transferred to VMF-211.

It was not until after the end of the war, in September 1945, that we learned that Boyington had become a prisoner of the Japanese and was still alive, and that George Ashmun had indeed been killed back on Jan. 3, 1944. Then after years of procrastination and, unhappily, excessive drinking, Boyington finally came out with his best-selling book, "Baa, Baa Black Sheep" in 1958. I last saw Pappy in 1959 in Cincinnati, Ohio, where he was kind enough to autograph his book for me. We reminisced about his last flight on Jan. 3, 1944.

I told him that I always felt that the rest of the flight had not given him sufficient support that ill-fated day. While Boyington did not seem to think there was a lack of support, there are some clues in his book that indicate that he did believe the rest of the Corsairs were slow to follow him down to sea level where he was finally shot down. For example, in his book his talks about the weather that day, "A few hazy clouds and cloud banks were hanging around—not much different from a lot of other days." It is my distinct recollection that there was an unusual number of cloud layers that did tend to disrupt the formations and the visual communication. He may have privately complained that we were slow to follow.

In my conversation with Pappy and from the remarkable story in his book this is what happened to him that sad day. Pappy and George Ashmun, being lowest in the formation, spotted about ten Japs just below them. Diving into this group they each shot down one plane. As they started back toward these Zeros, Pappy saw about twenty more planes above him which he thought were friendlies but were the enemy.

Ashmun and Boyington were weaving to cover each others tail. George was badly shot up, caught on fire and dove into the ocean. Pappy tried to escape but his gas tank erupted in flames. With an amazing effort he bailed out, pulled the rip-cord and landed in the water in one fell swoop. After floating around for an hour or two, he saw a submarine surface near him. They took him aboard and headed for Rabaul.

Pappy discovered that he was badly wounded. An ankle was shattered, there was a bullet through one calf, he had almost been scalped, and there were shrapnel wounds in his shoulder and groin. His wounds were not treated for ten days. He spent six weeks at Rabaul and six more at the Japanese base at Truk. He was beaten, starved, thirsty and had malaria. He was finally island-hopped to Japan. His weight dropped from 190 to 110 pounds. All this is recorded in his book, “Baa, Baa Black Sheep”.

Pappy writes in his book about how anxious he was to break Joe Foss' record of 26 victories as if it was a home run record or something like that, and not a matter of life and death. In his book he says several times that he was not interested in records but that it was the news media that kept pushing him. On the other hand, in several chapters of his book written about that period he seems to have been totally preoccupied with breaking Foss' record. It might take a psychologist to perceive what was really driving that gifted, complex man. He certainly was pushing things a bit too far. On that particular mission perhaps his precipitate dive from 20,000 feet down to sea level was ill advised. To us fighter pilots, altitude was always a precious commodity.

At that period of time the Marines (to say nothing of the Navy and Air Corps) were pouring an avalanche of fighter pilots into the Pacific. The military situation in the Rabaul area was not critical. So for one pilot to knock down two or three more Japanese planes was not all that urgent.

Much misinformation has been put out in the media about the Black Sheep squadron. They were not misfits, they were not always breaking the rules, they were not failures in life. They were more or less just like the other Marine Corps Fighter Pilots. If you look at the records, you will see that there was a constant transfer of pilots from one squadron to another. The average pilot that entered early in the war probably was in three or four squadrons. They ran the spectrum from hard-drinking, hell-for-leather types like Gregory Boyington to quiet, introspective, professorial types like George Ashmun.

The following are some cold, hard facts about the Black Sheep squadron as gleaned from the Marine Corps and National Archive records: Although VMF-214 had an excellent record in the war by shooting down 127 enemy aircraft, they were only seventh on the list of Marine Corps squadrons. VMF-121 had the great record of 208 planes downed, doing it while flying modest-performing Grumman F4F Wildcats. Having many hours in both the Wildcat and the Corsair, I can tell you the Corsair was vastly superior. VMF-221, which I was in for a short while, compiled a record of 185 victories. I was in VMF-211 for a long time. They knocked down 91 planes. Among the best 25 Marine Corps aces, the only Black Sheep pilot was Pappy Boyington at the top of everyone with 28 planes. Among the eleven Marine Corps pilots who were awarded the Congressional Medal of Honor, Boyington was the only Black Sheep member.

The Black Sheep squadron had an exceptional Intelligence Officer in Col. Frank E. Walton. During the war his record of the squadron activities was superior to any that I have read—and I have read the records of 15 to 20 squadrons. Possibly it was his accurate and descriptive writing that caused the newspapers of that day to begin to pay attention to Pappy Boyington and to the Black Sheep.

In Pappy's chapters concerning the week or two before he was finally shot down, he talks continually about his health, trying to break Foss' record, the media pressing him, but says not a word about the fate of many pilots in his squadron. For example, on December 23, just eleven days before Boyington's last flight, VMF-214 pilots First Lt. James Brubaker, First Lt. Bruce Ffoulkes (a good friend of mine from our days on Palmyra Island), and Major Pierre Carngey (Pappy's Executive Officer) were killed in action. On that same day Boyington shot down four planes over Rabaul. Then on December 28, Second Lt. Harry Bartl, First Lt. Don Moore, and Capt. Cameron Dustin also were killed in action. No mention of any of these six tragic deaths was mentioned by Pappy. It is easy to read into this that he pushed his men into dangerous situations while seeking fame and fortune for himself and did not care for his men, but probably just the opposite would be nearer the truth. He did care for his men and in his book the sadness of those events very likely led him to ignore the subject.

Gregory Boyington was a complex and gifted man. He was physically strong, well coordinated, had excellent eyesight, had great courage, inspired the pilots that flew with him, intelligent enough to get a degree in aeronautical engineering, and developed key friends in the Marine Corps.

But on the down side (by his own admission) he was personally irresponsible, always heavily in debt, not a family man, a pronounced alcoholic, involved in countless "barroom brawls", sought personal glory, made many enemies (being politically naive), and seemed to be accident prone. Like many a genius he was loved by many and hated by many. For my part, I know that if I were to meet up with a bunch of enemy planes, I would choose to have a fellow like Pappy with me rather than any other pilot I can think of with the possible exception of gentleman Joe Foss.

CHAPTER 37

"A Combat Mission"

by Wallace B.Thomson

On January 23, 1944 on the southwest coast of Bougainville in the Northern Solomon Islands, forty-eight F4U Corsairs, fighter planes attached to Marine Fighter Group Fourteen (MAG-14) coughed and sputtered as they were started one by one. We were poised to fly on a fighter sweep to the major Jap base at Rabaul on the island of New Britain 200 miles to the northwest across the Solomon Sea. Most of us were on Piva airstrip, the second of three runways built by the Sea-Bees since the invasion of Bougainville that began on November 1, 1943. Piva was a beautiful coral runway surrounded by coral taxi-ways, concrete plane revetments and towering jungle banyan trees.

Five miles away the Japs ringed the American invasion salient waiting to pounce if they could. The superb American Third Marine Division had carved out a semi-circle in the swamps, tangled undergrowth, rugged hills, mosquitoes, tropical downpours, and strange fevers that came with the Bougainville terrain. By now the front lines were manned by a U.S. Army division waiting for the inevitable Japanese counter-offensive.

One by one the Corsairs bumped their way onto the runway, their Pratt & Whitney engines roared, and then one by one our blue and white camouflaged craft with the famous inverted gull wings climbed out over the jungle in a gentle curve to the left. At least six Marine fighter squadrons were represented on this mission—each squadron contributing eight planes. All the fighters on this Rabaul fighter sweep were Corsairs. My plane was about in the middle of the pack and was followed by that of my faithful wingman, Ward Hower.

The flight leaders made a sweeping circle to the left around the Bougainville invasion pocket at Empress Augusta Bay, allowing all the planes to join up in formation by cutting across the circle. Below were the tumultuous, jungled mountains forming a hundred mile spine along the length of Bougainville. About 15 miles inland the active, 10,000 foot volcano, Mt. Bagana, put out its usual wisp of black smoke. As usual, one or two planes didn't make it, having mechanical, radio or sometimes even "psychological" problems. We headed northwest along the south coast of Bougainville then swung somewhat left toward Rabaul. There were Jap airstrips at Buka at the extreme northwest end of Bougainville. We wanted to stay well away from them, especially their coast-watchers, who could and probably did in this case radio an urgent warning to their comrades over at Rabaul.

Soon we were at 10,000 feet and shifted our super-chargers into LOW BLOWER. Most of us had been on oxygen right from the start. We were in our standard Thach Weave formation—planes flying in pairs, pairs flying in groups of four, and so on. The gray-green island of Bougainville with its smoking volcano gradually faded from view. On our left was the slate-blue Solomon Sea with its yawning New Britain Trench, an abyss going down to an incredible depth of 30,000 feet. To us, of course, 30,000 feet was no worse than 30 feet.

This was a fighter sweep. There were no dive bombers or torpedo bombers to escort. That very morning we had escorted SBD's and TBF's to Rabaul and had run into a flock of Zeros. Now we were coming back to challenge the Zeros once again. Only 20 days ago we had lost our controversial and peerless leader, Pappy Boyington, on just such a fighter sweep. On that grim day there were confusing layers of clouds that fouled up the mission, but today there was hardly a cloud in the tropical sky.

In a few minutes I charged my six .50 caliber Browning machine guns, three in each wing. Then, at a safe moment, I veered a little to the right and gave them a very short burst to see if they were working. The familiar "chug, chug. chug, chug" sound and curving tracers out ahead of me showed that the guns were working OK. I never had any gun trouble at all during the war.

The Corsairs had auxiliary gas tanks built into each wing near the wingtips which we all switched to soon after takeoff. Soon these ran dry and we all switched back to the main gas tank that held 230 gallons of high octane gas. To avoid a possible explosion in the empty wing-tip tanks during combat, we turned a valve that purged both wingtip tanks with carbon dioxide.

These early-model Corsairs had a plexiglas window at the bottom of the cockpit through which the pilot could (in theory) see downward. This window was usually smeared with oil and dirt so you couldn't see much. Today, however, I could see oil trickling back from the engine in a little greater quantity than usual. For the time being, I paid no attention to that.

At about 18,000 feet we switched to HIGH BLOWER which would allow us to climb up as high as the stratosphere. The entire formation continued to climb until we were a little above 25,000 feet. We headed up St. Georges Channel having the long and curving island of New Ireland on our right and New Britain with its port of Rabaul on the left. To save fuel we had been cruising at 1750 RPM and about 24 inches of manifold pressure. But now we were up to 2250 RPM and 32 inches. At this point I noticed a little more oil was flowing back from the engine over the window below me.

The formation was now north northwest of Rabaul so with the sun in the SW we had a good look at the fortress below. Simpson Harbor glistened in the sun. A few small ships—perhaps barges—were in the harbor. Rabaul had five airfields, the most important one being Lakunai right next to Simpson Harbor and the city proper. Suddenly I was alerted by huge clouds of dust arising from Lakunai and another field to the south which I judged to be Vunakanau. This was a signal that Zeros were taking off to meet our challenge. This was exactly the purpose of our fighter sweep—to tempt the Jap Zeros into the air and then shoot some of them down. The dust clouds were typical of Japanese air operations as the American coral airstrips and steel Marston mats had (with some notable exceptions) much less dust.

By now we were up to 2600 RPM and almost at full throttle, and probably not much below 400 MPH. The sky was clear in all directions and at that altitude appeared almost black. While the air temperature outside was probably close to thirty degrees below zero Fahrenheit, I was sweating from the intense sun rays and from tension.

At that point I was dismayed to see the oil pouring past the lower window in a thick, twisting river. As our whole fighter formation was making a giant circle to the left over Rabaul, a circle possibly 20 miles in diameter, I thought that Ward and I would be smart to throttle back a bit, cut across the circle, perhaps run into some climbing Zeros and, hopefully, join up with the rest of our Marines. So I cut back to 2300 RPM, reduced the throttle setting then headed across the circle. Ward followed without hesitation.

In a few minutes we were above the northeastern tip of New Britain and almost above the city of Rabaul itself. The river of oil slackened noticeably. Suddenly two planes appeared to the left and a few thousand feet below us. In the brilliant sunshine, the red meatballs on the wings were easily seen—they were two Jap Zeros, climbing in close formation, heading eastward.

I put my Corsair into a shallow dive and headed for the lead plane. Glancing around to my right I could see that Ward had also spotted them and was diving with me. The Jap wingman, also on the right, began acting nervously but the leader never budged as we closed in on them. The wingman rocked his wings then shot out a few hundred feet from his leader. Then he moved back in formation, then back out again, then back in again—all this in the space of a few seconds. Then to add to his bizarre display he flipped over on his back and flew in perfect formation.

All this time I was heading for the leader. When I was very close, maybe 60 to 80 yards, I opened fire with the six Brownings. He burst into a ball of flame almost at the moment I swept by his left wing-tip. I fancied that I could hear the explosion. My Corsair shuddered as I hit the blast and I also felt that something hit my plane. I remember being annoyed that most of my tracers went slightly left and that I should have trimmed my ship during the long high speed dive to avoid being in a skid.

Looking back I could see that Ward had shot down the wingman. But that Zero never did burn—it just went into a grotesque and jerky spin in the manner of a series of stalls.

Ward quickly joined up and we headed east. We had lost a lot of valuable altitude and we knew that Zeros could be all around us. However, as we continued east we saw none. The oil leak still looked bad. Also, looking out at my right wingtip I could see that somehow I had sustained damage out there. So after getting out of the Rabaul area we continued on back to the Piva airstrip at a low cruising power.

In about 45 minutes Empress Augusta Bay and the Torokina base came into view. My engine never gave any trouble from the loss of oil. The two coral runways along with the steel Marston mat of the Torokina strip were easy to spot in the green jungle of Bougainville.

After landing I found that I had indeed lost about half my oil supply. Years later after studying Marine Corps records, I noticed that there were quite a few engine failures during high performance combat missions, failures that often appeared to be caused by loss of engine oil. But my Pratt & Whitney functioned OK on that mission in spite of a heavy loss of oil. In four years of flying in World War II, I never had an engine failure.

Looking at the wing-tip damage, we judged that it was from exploding debris from the leading Jap Zero. Sometime later I found out that the plane I flew on that mission was one of the Corsairs manufactured by Goodyear and not Chance-Vought. Goodyear had just started producing the Corsair—their designation for that fighter being FG-1A and not F4U-1A. The severe oil leak was possibly caused by the usual bugs found in a production line that had just started.

Ward commented on the Zero I had shot down, "That ball of flame could have been seen for a hundred miles." We walked over to the mess hall. After little breakfast, no lunch, and two exciting missions to Rabaul, we were ravenous. Somehow being on oxygen all day always made me hungry—hungrier than at any other time in my life.

The fighter sweep was a success. As I recall, the Marines lost no Corsairs that afternoon and shot down ten or twelve Zeros. How could the Japanese live with such losses? The pressure we put on Rabaul during the past few weeks had been tremendous. As it turned out, the mighty base of Rabaul was bombed and strafed unmercifully while hundreds of Zero fighters were shot down by the Allied forces in late 1943 and early 1944.

CHAPTER 38

"Trouble at Green Island"

by Wallace B. Thomson

When I first arrived at "Buttons" or Turtle Bay on the east coast of Espiritu Santo in the New Hebrides islands in October 1943—that great Marine Corps base that fed Marine aviation power into the brutal campaigns in the Solomon Islands—we would hear all kinds of stories of what the war was like.

In one story, a Marine flier, whose name and organization I never learned, was flying in formation over the remote interior of Espiritu. This was incredibly rugged country and supposedly infested with head-hunters. Suddenly, without warning, one wing of this poor fellow's plane folded and he gyrated wildly into the jungle below. Since his death was certain, the difficulty in reaching him was so great, and our resources for doing so were so limited at that time, no recovery effort was mounted. After the war ended, I'm sure people went in there to find his remains. Since very many of the planes the Marines flew were designed for aircraft carrier operation, they had wings that could be folded. In rare instances they would fold accidentally. Stay tuned!

Green Island, also called Nissan Island, is an atoll that lies about half way between Bougainville and New Ireland—about 75 miles from each. Green Island has an oval shape being eight miles across in the north-south direction and four miles east to west. As in almost all atolls, at the center was a large lagoon. The land area formed a narrow ring, averaging about three quarters of a mile from ocean to lagoon.

On February 15, 1944, 5800 New Zealand troops, protected by Marine Corps fighters, invaded Green. After three days they had full control of the area. There were about 100 Japanese on this island of which number 30 were killed. The 33rd, 37th, and 93rd Seabees worked hard to construct a base there. By March 15 the airstrip was fully operational.

On March 20 the squadron I was in, VMF-211, the "Wake Avengers", was based at the new airstrip on Green Island. About fourteen pilots of the "Black Sheep" squadron were now with the "Wake Avengers".

On the next day, March 21, I was at the squadron ready hut prepared for whatever missions were planned. Our squadron ground crew was in the process of relieving the crew of the previous squadron. I was informed that the plane I would be flying for the next few weeks would be number 027 which number comes from the Bureau of Aeronautics plane number of F4U-1 No. 56027.

One of the ground crew (I don't recall whether it was one of ours or one of the other squadron's men) came up to me to talk about the plane. He said that the plane had been in an accident with the previous squadron and as a consequence the left wing had to be replaced. He said that the new wing had been installed by the previous ground crew and that the plane should be test hopped before going out on a mission. As I was the Engineering Officer of our squadron and as the plane was one I would be flying regularly, it was only natural that I would be doing the test flight.

Now the F4U Corsair was primarily designed as an aircraft carrier fighter plane and hence had wings that could be folded up to allow the craft to be moved from deck to deck on a carrier elevator as well as to provide room for more planes when placed side by side on any given deck. In this particular instance, the installation of a new wing meant that part of the wing that extended from the joint where the folding occurred out to the wing tip. What ever damage had occurred, and I never learned what it was, had been in that outer part of the left wing. The Corsair plane had that famous inverted gull wing. These wings folded at about the lowest point of the curvature.

During carrier operations the Corsair wings were spread and folded on every flight. All operations with which I was involved were land-based ones where we never folded the wings. In two plus years of flying the Corsair I never once folded the wings nor did my fellow pilots. Nevertheless we all were familiar with wing folding, with the two controls in the cockpit: one that placed the wings in "Spread" or "Neutral" or "Fold" mode, and the second which would "Lock" or "Unlock" the wing hinge pins. When the wings were spread as they always were for us the first lever was always moved aft in the "Spread" position while the second was always placed forward in the "Lock" position.

The wing hinge pin that, when in place, prevented the wing from folding in flight, was a solid cylinder a little over an inch in diameter and made of hardened steel. To indicate that the hinge pins were in place or "home" as they called it, there were indicator doors at the upper surface of the wing at the wing joint. These doors, measuring perhaps three inches wide and six inches long, lay flat, flush with the wing surface when the wings were "Spread". If the hinge pins were not in place and the wing could fold, the little doors stood up on edge and one could see the red paint of their under surface.

If there was any doubt about whether a wing was safely locked or not there were four things one could do to check things out. So when I went out to Corsair 027 that day, I climbed into the cockpit and checked the wing folding levers. They were in the correct position but seemed to be stiff rather than resting easily in place. Then I glanced at the indicator doors on both sides: they appeared to be flush with the wing upper surfaces. But afterward I wished I had taken a harder look at them. Going down to the landing gear I located the cables connected to the folding mechanisms. The cables were supposed to be somewhat loose—and they were. Finally, following our pilot's handbook instruction, I went to each wingtip and reaching up shook each one as hard as I could. Both wings seemed to be solid, so as far as these checks went, everything was normal.

I prepared to go on the test-hop and soon was moving down the coral taxi-way. The tower surprised me a little by directing me to take off from the north end of the airstrip. I could clearly see that the various wind-socks were pointing south so I would be taking off downwind. For a fighter plane with high performance this was no big deal but it still was unusual. Arriving at the north end I could see the tower's reason for sending me there. Gathered at the extreme south end of the runway, perhaps 3000 feet away were, about 30 P-40 "Warhawks". Each one carried below its fuselage a 500-pound cluster of thermite bombs—bombs impregnated with magnesium to start intense fires. They were getting ready to pay Rabaul a hot visit. Their props were turning over and they were only minutes from take off.

The tower wanted to get me out of the way first so gave me a green light. So I gave my ship the throttle and headed down the airstrip expecting that I would be several hundred feet above the P40s by the time I reached them.

My Corsair had picked up speed to about 60 or 70 knots, close to take off speed, and was still on the deck when something on the left caught my eye. I was horrified to see that the indicator door at the left wing joint had popped up. The wing was unlocked! An instant later the left wing shot upward so that the wingtip was straight above me. It hit the top position with a bang, went all the way down with another bang, then up with yet another bang—the damned thing was flapping like a wounded bird! I cut the throttle instantly and applied the brakes, but the weird aerodynamic forces on the wing made the Corsair snake violently down the runway.

My Corsair raced toward the P-40s and their thermite bombs. With all my frantic braking I was never going to stop in time. If I had any thoughts of a hellish finish I don't recall them now. But as I flapped on down toward the P-40s they saw me coming. Some went right, some went left—parting like the Red Sea in Moses' time. Having no control at all I still missed them all, went off the end of the airstrip and into a swamp filled with tree stumps.

I was unhurt but livid. Two crash trucks were there at once. An ambulance crew checked me out. Damage to the Corsair? I had knocked the tail wheel off the plane. They got me back to the squadron ready hut. Nobody there had noticed anything unusual going on. Who had botched the wing installation? The responsible squadron was leaving and couldn't be bothered. I tried to talk to the skipper, Tom Murto, but he was too busy moving into Green Island to worry about a minor accident that none of the fellows had even seen. The flight did not even appear in my log book for the simple reason that I had never left the ground!

I could never figure out, after checking all four items in the Corsair pilot's manual, why the wing could appear to be locked in place but obviously was not. Somehow, the hinge pin was just slightly in place and the levers in the cockpit were forced.

What about Corsair No. 027? Well, our ground crew repaired the tail wheel, reinstalled the left wing—this time correctly—and the plane was back in the air in a few days. My log book shows me flying that Corsair six days after the folding wing affair. Within the next few weeks I flew 027 eight times: escorted dive bombers to Rabaul, escorted dive bombers all the way to Kavieng, New Ireland, patrolled Green Island (almost like a day off), covered a naval task force, made a barge sweep to Cape Lambert, escorted B-25s to Rabaul, and several similar missions. So old No. 027 wasn't so bad after all. But my mind sometimes goes back to that unfortunate Marine flying over the rugged interior of Espiritu whose wing suddenly folded, plunging him into the remote headhunter country of that New Hebrides island.

“Pappy” Boyington in F4U-1A cockpit, Fall, 1943
(Photo courtesy Wally Thomson)

F4U-1 Corsair wreck at Torokina, Bougainville, December 11, 1943, before the runway was fully operational.
(Photo courtesy Wally Thomson)

PART VII

by John R. "Jack" Eckstein, CAPT USN (Ret.)

I was born in Massillon, OH, the eldest of four children. During WWII I developed a great desire to fly so I got a job at the local grass field and worked for my flight instruction, eventually soloing in a Taylorcraft L-2 on 17 August 1945. During high school summer vacations in '43 and '44 I also worked at Goodyear Aircraft Corporation which was building the FG-1 Corsair, 24 planes a day on three shifts, seven days a week.

I took Aeronautics in my junior year of high school, and became an Air Raid Warden even though I was supposedly too young for that. My buddy and I headed up the Air Scouts and the Civil Air Patrol Cadets as the older fellows entered service. Anything and everything I could do to prepare.

I joined the Navy in April '46 as a "Combat Aircrewman" designee because Flight Training was all but completely closed down following the end of the war. While in Aerial Photography school in Pensacola (as close as I could get) a program to accept a few Navy and Marine enlisted men for Flight Training was set up. After rigorous interviews, testing and screening, 500 of us were sent to Great Lakes for further screening and preparation for college.. It was tougher than "Boot Camp" but in the end, just thirty-three of the 500 made it through.

In late August of '47 we were sent home to attend the college of our choice for two years. The unrest in Korea shortened my time in the pre-engineering curriculum at John Carroll University in Cleveland, OH, and I reported to Pensacola as an Aviation Midshipman to begin flight training in February 1949. I received my wings on 16 August 1950, about two months after the Korean War began. Later on I completed my BS degree at the Navy Postgraduate School in Monterey, CA, finishing in '61.

My Navy career spanned thirty years. I retired as Captain, USN in November '75. Commands held : CO VAW113 in USS Constellation (CVA-64) deployed to the Gulf of Tonkin during the Vietnam action, CO RVAW 110, the VAW training squadron, and Commander Carrier Airborne Early Warning Wing 12 in Norfolk, VA.

Awards included the Bronze Star, five Air Medals, the Navy Commendation Medal, and the Philippine Presidential Unit Citation for Typhoon relief while serving as XO of USS Okinawa (LPH-3), along with the usual theater ribbons and Vietnamese awards.

CHAPTER 39

"Going to Japan"

by John R."Jack"Eckstein, CAPT, USN (Ret.)

Right after earning our aviator's wings at Pensacola, we all went on a short leave to our homes. In late August, 1950, Neil Armstrong, fated to trod the moon, Herb Graham, and I—all still Aviation Midshipmen—drove out to our first duty station, FASRON 7, at NAS North Island, CA, in my '50 Chevy. After a short time there and no flying, we were told that the USS Leyte (CV-32) was making a high speed run from Norfolk to pick up Air Group Three and 33 replacement aircraft in San Diego, then to high-tail it to WestPac (Western Pacific) to join the fray in Korea.

A number of us, some still Aviation Midshipmen (even though we had our Naval Aviator's gold wings!) were assigned TAD (temporary additional duty) to Air Group Three as ferry pilots. I was assigned to VF-33 in Ready Room Two well aft and below the hangar deck, and to the junior officer's bunkroom well forward and just beneath the flight deck. All the bunks were occupied except for one at the top of a three-decker, filled with everybody's baggage. Mine.

The morning after we got under way "General Quarters" was sounded at about 0500. I came straight up in the bunk out of a deep sleep. Then I saw lots of stars and felt a great explosion in my forehead as I rammed into an overhead steel beam supporting the flight deck. In pain and confusion I stumbled frantically aft and down to my battle station in Ready Three trying to beat the setting of Condition Zebra when all water tight doors and hatches would be sealed for battle.

(Condition Zebra is the most watertight condition which can be achieved and is standard fare when preparing for battle. All watertight doors and hatches are battened down and it becomes almost impossible to move from one's battle station. Which in my case was Ready Three! Normal cruising condition is condition Yoke. X-Ray is set in port during working hours to allow maximum access to working spaces. Yoke is reset at conclusion of working hours.)

When we arrived off Pearl Harbor on 23 September, 1950, the air group as well as the replacement aircraft, 30 Corsairs and 3 Avengers—were to be flown off . I hadn't flown an aircraft since final CARQUAL (carrier qualification) on 10 August at Pensacola, and I felt a bit anxious because the Corsair was a considered a real handful, especially for a rather green aviator who hadn't flown for over a month. We manned our aircraft and I was directed to the port catapult for my first ever CAT shot! All went well until I tossed the catapult officer a salute and put my head firmly back against the headrest while crushing the catapult grip and throttle in near panic. An instant before the firing stroke I realized I was STILL STANDING on the brakes! I got my feet off just in time!

We spent three days at Barber's Point during the airgroup workup, the TAD crew and some of the airgroup handling liaison ashore, and then departed at 25 knots for Japan. Arriving in Tokyo Bay I got ready for my second cat shot and a big formation flight proudly announcing our arrival at Kisarazu Air Force Base on the north side of Tokyo Bay. All went well this time and I joined up for the short hop to the base and landing.

While I was walking into base ops with my section leader, a marine Major who was a member of the airgroup, one of the hot squadron fighter pilots approached the runway overhead, executed a beautiful hot break over the runway and pulled a tight pattern to touchdown. But he forgot to lower his landing gear and slid unceremoniously to a stop in view of the whole world, especially the Air Force! I said a silent prayer of thanks that it wasn't this little ol' Midshipman who was not yet quite dry behind the ears!

I arrived in Japan with a total of nine dollars in my pocket. As you will recall, a Midshipman earned $78.00 base pay and 50% or $39.00 flight pay at that time. No big spender on the Ginza that night! Neil and Herb didn't make that trip but they eventually deployed with Air Group Five and I volunteered for a squadron flying Corsairs at Alameda, CA.

CHAPTER 40

"Alameda"

by John R. "Jack" Eckstein

Not long after Carl Quitmeyer and I checked into VF-194 in Alameda in early 1951, a fellow whose name I can't recall checked in from the Training Command. He was scheduled for a familiarization flight with a lad named Larry. Larry was the type who was always "alone" up there; he was a nice guy but he never gave a wingman a second thought! On that fam flight he led the new arrival through a real aerobatic wringer. Larry took him into a loop from which the "victim" spun out inverted into a series of 5 spins, the first three inverted. The new gent spun down from 19,000' to about 3000' before recovering. When they got back he was soaked with sweat, I think it was sweat, and was white as a sheet. Still shaking! Maintenance found six rivets sheared in the right rudder pedal. Both feet pushing on that mother!

While we were in El Centro in May of '51 we arose at 0400 and flew until about noon because the planes became too hot to touch in that desert heat. We had gunnery practice on the morning of 7 May following some formation work and I was scheduled to tow the target banner.

At the Skipper's order I broke from the formation and landed to pick up the tow which was laid out alongside the runway. After hookup I lined up for takeoff and added throttle smartly. All of a sudden I was headed left off the runway and standing on my right rudder pedal with my heart in my mouth, cranking furiously on the rudder trim tab with my left hand! I straightened out straddling the left side of the strip and raised an immense cloud of dust as I managed to get airborne without hitting anything! What were those takeoff rudder settings? About 6 degrees right rudder? Of course, I still had my rudder trimmed for cruise. About 4 degrees left or so. So much for a thorough review of the takeoff check list! Sometimes I'm just amazed we survived our naive bravado.

When Carl and I were in VF194 at Alameda flying the "Ensign Eliminator," the squadron was requested to do a fly-over of the USS General Breckenridge as it entered San Francisco Bay with the first Marines returning from combat in Korea. I'm almost certain that these Marines were the ones who had fought their way out of North Korea after the Chinese came across the Yalu River in December '50/January '51. They brought out their dead and wounded, and their equipment, leaving nothing behind—heroes all!

We had 24 F4U-4's in the squadron, and I think we had about 16 going out to greet the ship. Our Skipper, LCDR Robert S. Schreiber, would lead the squadron in four four-plane divisions, we would fly low to the mouth of the bay and reverse course over the Golden Gate Bridge to swoop low and pass over the ship at mast top level.

Because of our altitude the usual step down between aircraft and divisions was to be rather flattened. We manned aircraft and taxied for takeoff in assigned order. I was to be in the third division. When I ran my magneto check the engine ran very rough on one bank and I knew I couldn't make a safe take off.

I hurriedly taxied back to the flight line, grabbed my chute, and left the plane intending to return to the line shack for another plane. But I didn't want to miss the excitement so I asked the plane captain if there was another "up" aircraft available. He said there was and I jumped in and taxied back to the runway for warmup (minimal) and a mag check. This time it ran smoothly so I took off and raced to join up.

When I reached the formation it had already begun its turn and descent. As I slid in to join my division I couldn't reach my spot and used hand signals to get one of the planes to slide accross and give me his spot. With San Francisco's skyline so close below us he was very reluctant, but finally moved over and I managed to be in place as we buzzed the Marines, who were all facing skyward to watch us.

The newsreels later showed the approaching squadron with one late comer catching up. Upon return, the squadron maintenance officer took a very large bite out of my behind for my foolish actions. Fortunately, our skipper was a WWII wild man and he decide not to ground me.

PART VIII

Miscellaneous Tales of Trial & Terror!

These are stories from some former VF-14 "Tophatter" Corsair pilots I know. All these "tales" occurred after I left the squadron, so I was not able to get the actual accident reports. . .

"Cookie" Cleland tells us how his flight leader got killed flying into the water while on instruments—and how "Cookie," flying formation on him in heavy clouds, pulled away just in time!

Gene Hendrix describes his concern about getting 1000 accident-free hours when he had a barrier crash.

And Randy Moore found out Corsairs don't float!

CHAPTER 41

"COOKIE" CLELAND

by Fred "Crash" Blechman

William L.Cleland got his squadron nickname of "Cookie" because at the time Cook Cleland was a well-known air race pilot and Naval Aviator. Originally from Franklin, Tennessee, Cookie is now retired with his wife, Jean, in Brentwood, Tennessee.

Cookie went through Navy flight training as an Aviation Midshipman, and got his Navy "Wings of Gold" in April 1950 after flying F4U-4 Corsairs in advanced training. He joined VF-14 on June 23, 1950 and his first F4U-5 flight was on July 31, 1950. His logbook tallies 224 carrier landings, of which eight were (ugh!) NIGHT carrier landings.

HERB LANDIS LOST AT SEA

Of all the hundreds of flights Cookie Cleland made in Corsairs, one that still stands out 45 years later was a deck-launch from the carrier USS Franklin D.Roosevelt (CVB-42) on October 6, 1953, during Operation Weldfast. On this pre-dawn launch, Cookie was scheduled to fly as a Tactical Air Observer of a two-plane CAP (Combat Air Patrol) flight by LT Herb Landess and a wingman.

As he recalls, "There was about a 300-foot overcast. Herb and his wingman were to fly over Greece, and I was supposed to get up there, circle around, and report on the action. But Herb's wingman downed his airplane for a mechanical problem, so I intended after takeoff to join on Herb's wing.

"First Herb took off, then I took off and both of us climbed up through the overcast separately, on instruments, of course. It was several thousand feet thick. When we got above the clouds, I joined on Herb as his wingman, and we began flying over Greece northwest to the mission area.

"It was dark as it could be, and I was flying close on his wing. We had been airborne about a half-hour when he inadvertently flew us into a thunderstorm. You can imagine, when you're flying wing on somebody in real rough, cloudy—you know, in the soup—you're sitting up there just as tight as you can be, so you can't refer to your instruments at all."

Formation flying is a learned art. You quickly learn that you can't look away from the plane you're formed on for an instant, or you miss the tiny stick and rudder changes that are constantly—and unconsciously—necessary to keep you in position. And the closer you are to the plane you are flying formation with, the more immediately obvious are the changes in bearing and distance that require your corrections. Of course, where visibility is lost in heavy clouds, you must either break away and fly on instruments, or trust that the plane you are flying on is on instruments and flying a safe path—usually level flight, a slight climb, a slight descent, or a slight turn. This is no time for drastic maneuvers!

Cookie continues, "I suddenly realized I was pulling Gs to stay with him. We were practicing radio silence, so I always flew with my microphone up over my helmet—I didn't have it down where I'd be tempted to use it. So I didn't try to call Herb. I pulled away and went on instruments, and found we were in a graveyard spiral—he had us in about a 100-degree bank, and pulling as hard as he could!"

This sounds to me like a pure case of vertigo—pilot disorientation. I've had it myself, when in a cloud and trying to catch up with the Corsair ahead. I was looking through the windshield instead of my instrument panel, and would have sworn I was in a normal climb. When the engine started whining, and the wind past the canopy began whistling, I looked at my instrument panel and found myself in a steep left, descending turn—a "graveyard spiral"—apparently a common disorientation where engine torque rolls the airplane to the left if controls are not perfectly trimmed.

Unfortunately, recognizing you are in a graveyard spiral can lead to further difficulty if your initial reaction is to pull out of the descent by pulling back on the stick. This tightens the turn, possibly into a high-speed stall! The proper procedure is to FIRST level the wings, hoping your descent doesn't slam you into the ground, and THEN pull up the nose. . .

Cookie did it right. "You know, you roll your wings level and pull out. I got out at about 500 feet above the water, still in the soup. Since it was still dark, I had to wait about a half an hour before we could start looking for Herb. My total time on that flight was 5.8 hours, most of it looking for him. One oil slick was found by a destroyer; that was probably where Herb went in. We figure he must have had a heart attack or something, because Herb was a good instrument pilot."

BARRIER CRASHES

Cookie also had the thrill of a couple of barrier contacts—a common experience for those of us flying the long-nose Corsairs from small straight-deck carriers in those days. Actually, the second "crash" was so minor—slight prop contact with the barrier cable—that it did not even warrant an accident report.

There were basically three types of U.S.aircraft carriers in the late '40s and early '50s: CVEs, about 525 feet long, with eight arresting wires; CVLs, about 650 feet long with nine arresting wires; and CVs, about 850 feet long with (as I recall) 13 arresting wires.

Later on there were CVBs, CVAs, and CVNs with angled decks and only four arresting wires—but you could go around again (called a "bolter") if you missed a wire. We did not have the luxury of a bolter. We landed with four sets of barrier cables across the deck forward of the last wire. The Corsair was heavy enough that if your tailhook caught the last wire, it would stretch out just far enough for your big prop to bend a couple of blades on the first barrier cable!

The F4U-5 Cookie was flying contacted the barrier twice while in VF-14 on the light carrier USS WRIGHT (CVL-49) on a Mediterranean Cruise—January 30, 1951 (his 38th carrier landing, aircraft #121947, side number 404), and March 1, 1951 (his 44th carrier landing, aircraft #121860). Both were last-wire barrier engagements, although the second one was so minor they only had to use a file on a couple of the prop blades.

But the second one may have caused some undetected engine damage, because about six months later Cookie was flying the same aircraft when, as he relates, "I was flying that same plane, #121860, on a practice dive-bombing run off of Jacksonville, when the engine kind of exploded on me in the dive. I don't know what happened to it. It was running awful rough as I was flying back to the field about five or six miles away. I landed and the engine quit as I was rolling out!"

EPILOGUE

After fleet time, and while still on active duty, Cookie became an advanced flight training instructor in the Naval Air Training Command. He instructed in the F6F Hellcat, T-28 Trojan, F9F Panther and TV-2 Shooting Star.

After active duty, Cookie joined the Naval Reserve in Memphis, flying the F9F-6 Cougar and FJ-4 Fury, then went to the Atlanta Naval Reserve as an instrument flight instructor in a jet F-8 Crusader squadron. He retired from the Navy as a Lieutenant Commander (LCDR).

As a civilian, he became a rated multi-engine and instrument flight instructor, doing private instruction. He also pursued a career in the chemical industry in both engineering and sales.

CHAPTER 42

GENE HENDRIX

by Fred "Crash" Blechman

Originally from Birmingham, Alabama, Gene is now retired with his wife, Mickie, in Fort Walton Beach, Florida.

Gene was an enlisted Navy seaman when he was accepted for Navy flight training as a NavCad, and got his Navy "Wings of Gold" in September 1951 after flying F6F Hellcats in advanced training. He joined VF-14 in November, 1951 as the first Ensign to join the squadron in a year. (I was the last Ensign before Gene.) His first F4U-5 flight was on December 5, 1951. His logbook tallies 169 carrier landings, including one at night.

FIRST CORSAIR FLIGHT

Gene says his first Corsair flight was the most interesting flight he ever had! "I downed three aircraft before I got airborne. On the first plane, during my preflight walking around the plane, I moved the horizontal stabilizer up and down and the damn thing squeaked! I asked why it squeaked. Nobody knew. I said there's nothing in an airplane that should squeak. I downed the plane. That's what I put down—it squeaked.

"I go down and get another and down it for the same damn reason. I go get a third one, and it had an ammunition link hung up in the wing-fold mechanism, so I didn't get a safety indicator for the wing-fold. By this time everyone is panicked. So I finally get airborne, flying with LT Emmett "Bob" Brown in another Corsair alongside, checking me out. We had only been up about 30 minutes, flying down the St.Johns River, when I added power. All of a sudden, Bob calls on the radio and says, 'Turn around, let's go back. Pickle off your belly tank over the river and then go back and land. . . ' I didn't know it, but when I had added power smoke was coming out. He thought after me downing three planes and then seeing smoke coming out on this one, that I was just having a bad damn day. So on my first Corsair flight, I had downed three aircraft, dropped a belly tank, and had only flown eight-tenths of an hour!

"When they checked for the squeaking on the horizontal stabilizers of two of the planes I downed, they found a fitting had been installed upside down on both planes!"

ONLY ONE ACCIDENT

Although Gene spent 20 years flying several different types of Navy aircraft on active duty and in the Navy Reserve (including jets), he only recalls having one accident.

It was August 25, 1953. Gene was deck-launched in an F4U-5 (#121811) from the carrier USS Franklin D.Roosevelt (CVB-42) while on a Mediterranean cruise. The 2.2-hour flight off of Genoa, Italy, with three other Corsairs was normal up until the landing. An Ensign at the time, he was the last to land in his flight. While going through all the intricacies of preparing the plane for landing—wheels, hook, wing flaps, cowl flaps, prop, power, setting the abeam position, slowing down the plane, turning toward the ship—his thoughts wandered. He had an accident-free record all the way through flight training and his time in the fleet, and wanted to maintain that record. With this flight he would getting close to 1000 hours of pilot time.

"The landing was normal. I caught a wire, but instead of holding the stick back to keep the tail down during the time the arresting wire was pulling out, I released the stick too soon. I was so relieved that I had made it to 1000-hours accident-free that I just sat there and let the stick go forward! I nosed over just enough to have six inches of the prop hit the deck, bending all four blades. After that, everything was just great—I never had another accident."

As an aside, another pilot with Gene in VF-14 was Gordon Flynn. "Gordon and I knew each other as babies—back to about five years old! We lived across the street from each other in Ensley, which is part of Birmingham." Gordon, and his wife Mamie, now live in Trussville, Alabama.

ONE-EYED PILOT

Gene recalled the VF-14 cruise to the Mediterranean on the USS Wasp (CV-18) from May 24, 1952 to October 10, 1952. That was the cruise where VF-14's Commanding Officer, LCDR "Jack" Kennedy—unbeknownst to anyone but himself—was flying F4U-5s off the carrier with one eye! As Gene remembers, "Within days after we got back to port at the end of the cruise, he went into the hospital and had one of his eyes removed! It was cancerous! I was shocked that he obviously had been flying—probably the last six months—flying off the carrier with one eye. That's got to be one of the most interesting stories in the world. How he did it just blows my mind every time I think of it. He apparently did it, and did it well."

That was also the cruise where LTJG Lawrence F.Emigholz, Jr., of sister-squadron VF-13, was killed on July 15 while attempting to land his F4U-5 aboard the Wasp. When he received a wave-off he added power too quickly and the plane torque-rolled to inverted when he hit the water.

NO THROTTLE CONTROL!

Another flight Gene remembers well was the time his throttle linkage broke! "We were flying in the Med off the Wasp. We were coming back to land when at one point I added power—and it didn't work! I had to drop out of formation, since I couldn't maintain airspeed. No matter how I moved the throttle, nothing changed. I could change the prop pitch, but not the fuel flow. I dropped back and they let everybody land. I was flying with "Hoppy" (LT Jesse Hopkins) and his last comment to me was, 'Let me get out of the airplane before you land.'"

Hoppy was concerned that Gene's Corsair, with so little control, would jump a barrier and land on top of him as Hoppy taxied and parked forward.

"I didn't know this at the time, but on a previous VF-14 cruise when the same thing—a broken throttle linkage—happened to the Skipper at that time, they diverted him to a land base. This time they didn't have this alternative, since we were not near enough to a land base.

"When I dropped the wheels and flaps, the plane was flying so slow I wasn't sure if it would even fly. I couldn't give up anything, or I was lost for good. I couldn't get too low, there was no way I could get the nose up without losing airspeed. I was holding 30-inches of manifold pressure, just enough to make a real wide approach, maintaining airspeed, and never gave it a chance to get low. When I got the 'cut' from the LSO (Landing signal Officer), I killed the engine by turning the ignition switch from BOTH magnetos to OFF!"

FLYING CORSAIRS

When I asked Gene if he recalled any other "tales of trial and terror" when flying the F4U-5 Corsair, his reply was, "All the damn time! As long as I flew the -5 Corsair, I never trusted the airplane."

He told me about the time the left inner flap would not come down during the turn to the downwind for an approach to a carrier landing. He wondered why it turned left so easily (the right inner flap WAS down, tending to lift the right wing). When he had trouble leveling the wings, he spotted the left inner flap still up! He aborted the landing approach and went to altitude to try to get all the flap down. The left inner flap would just not go down, which would make the plane uncontrollable in the final approach to the carrier. The carrier was not within range of a regular airfield, so he made a flaps-up carrier landing, something we never trained for!

He continued, "In the Navy Reserve in Birmingham we had F4U-4s—which were okay compared to the -5s—then F4U-5s, then we got ADs. I had only flown the AD two hours, and I liked the AD better than I ever liked any Corsair."

EPILOGUE

After five and a half years of active duty, Gene joined the Navy Reserve in Birmingham and then New Orleans, and flew the F4U-4 and F4U-5 Corsair, the AD Skyraider, the F9F Panthers and Cougers, the FJ-2 Fury, and the A4 Skyhawk. His favorite plane from the safety standpoint was the AD, but his next choice was the F4U-4—NOT the F4U-5. As Gene put it, "The F4U-4 flew better, had lower fuel consumption, it landed better, it was lighter and easier to handle. The nose on the F4U-5 was heavier than a son-of-a-gun."

Gene retired from the Navy in 1968 as a Lieutenant Commander (LCDR). In civilian life he followed a career as an electronics engineer from 1957 until full retirement in 1988.

CHAPTER 43

"Corsairs Don't Float!"

by Randy Moore

LTJG Randolph "Randy" Moore is originally from Royston, Georgia, and is now retired with his wife, Gwen, in Snellville, Georgia. Randy entered Navy flight training directly from college as an Ensign, and earned his "Wings of Gold" in July 7, 1950 after flying F4U-4 Corsairs in advanced training.

He was immediately assigned to VF-14, arriving in July, 1950, and had his first F4U-5 flight on August 5, 1950. His logbooks records 428 day and 43 NIGHT carrier landings in various aircraft.

Here's a story of one very short flight, as related by Randy:

CORSAIRS DON'T FLOAT!

On July 18, 1953, just four days before my 26th birthday and only 18 days before the birth of my third daughter, I dropped off the flight deck of the USS Franklin D.Roosevelt (CVB-42) into the sunlit blue water of the Mediterranean Sea, firmly strapped into the seat of an F4U-5 Corsair belonging to Fighter Squadron 14 (VF-14).

Believe me, Corsairs will not float! The remains lie several thousand feet under the surface at 36-degrees 11-minutes N and 23-degrees 36-minutes E. I imagine that my old navigation board is still in the instrument panel slot, complete with the details of our flight plan for a simulated close air support mission in Greece.

The size of the FDR permitted a double line of planes to make take-off runs on deck—left lane to the right corner of the bow, and right lane to the left side of the bow. My run was from right to left. I applied power up to about 40 inches of manifold pressure with the brakes locked, and held the stick all the way back to keep the tail down.

After checking my instruments and saluting the Flight Deck Officer, he gave me the launch sign. I released the brakes, gave the throttle the final push to full power, and started my roll. The squadron mechanic who had helped check the plane said the engine sounded good. To me it sounded good. But as I rolled forward and the automatic turbo cut in to give me around 61 inches of manifold pressure, I realized that the lack of pressure pushing me against the back of the seat indicated the presence of a serious problem. I wasn't accelerating quickly enough!

I knew that I could not stop the plane on the remainder of the deck, and that there would not be enough speed to fly when I got to the end of the deck. As I went off the bow I took a hasty swipe at the gear handle to raise the wheels (it's bad enough to make a water landing in a radial-engine airplane, but wheels down make a flip-over even more likely!)—but the wheels were not up by the time I contacted the water.

Also, I did not have time to release the external fuel tank. The plane contacted the water to the port side of the oncoming carrier at about a 30-degree nose down attitude, the fuel tank split, and fire erupted all around me.

I definitely remember thinking that I may not see my soon-to-be-born child. As the plane sank I was pulled down through the small flames on the water that was flowing into the cockpit. By now the heavy engine had pulled the plane into a straight down angle. Somewhere under water I released my seat belt and pushed up from the seat (which at that angle meant parallel to the surface) and I was free of the plane. When I surfaced there were flames all around me with much black smoke.

Thankfully we had launched into a surface wind of about 20

knots, and the burning gasoline was being blown away from the direction I knew I must swim. However the rescue helicopter dropped the old "horse collar" rescue device into the area of the sea that was covered with flames! The rescue pilot finally saw me signaling for him to drop the collar further upwind. (I may be stupid but not enough to go back into those flames to get into that thing.) He then dropped it out of the flames and close enough to me to use. I was picked up and delivered to the flight deck, then to sick bay for a check-up. No injuries, but completely shaken up.

When I went to the barber about three days later he did not know that I had "crashed and burned." He did comment that my eyebrows, and the hair below where my helmet ended, appeared to be singed. This occurred when I was pulled through the flames on the water.

The cause of the low airspeed was determined to be the failure of the "take-off door" not closing properly to route turbo-pressurized air to the carburetor. Although the instrument read 61+ inches of manifold pressure, it did not get to the right place.

Jocelyn was born on August 5, 1953. I first saw her on December 3, 1953, gratefully and with tears in my eyes. . .

EPILOGUE

Randy left VF-14 in 1954 after three Mediterranean cruises (USS Wright in 1951, USS Wasp in 1952, and USS FDR in 1953), and numerous carrier deployments during Caribbean exercises. During the last part of his time with VF-14 he transitioned to the F3D Sky Knight twin-engine jet night fighter. Subsequently he was flying AD-5Q's with VAW-33 and was O-in-C of a detachment aboard the Saratoga. Later in his Navy career he was Commanding Officer of BNAO (Basic Naval Aviation Officer) training at Pensacola, which used T1A Shooting Stars, and SNB Navigators. That unit later became known as VT-10 for Naval Flight Officer training.

He retired from the Navy in 1969 as a Commander (CDR) and as a civilian developed an accounting and insurance firm. Now retired he manages a family farm where he has constructed a private RV campground for their friends.

EPILOGUE

"Air Combat U.S.A —Fighter-Pilot for a Day!"

by Fred "Crash" Blechman

(NOTE: This was written when I was 64 years old, in 1991. When I came down from this six-dogfight challenge I found my left eye flashing—perhaps from pulling sustained Gs often during the flight. The next day an ophtholmologist used a laser beam to spot-weld my left retina in 170 places, saving my eyesight in that eye! After that flight I decided to quit aerobatics, and went into flying gliders, and bumming rides in ultralights, homebuilts, and light planes. . .)

"Tally! Tally! He's low, he's low!" crackled in my headphones just as I spotted the bogey closing rapidly from the opposite direction and below my 5500 foot altitude. "Fight's on! Fight's on!" someone shouted as the red-trimmed silver Marchetti streaked by about 500 feet off my left wing and immediately rolled into a tight turn toward my tail. I was going too fast so I pulled the nose up sharply to drop my speed for better cornering. Banking to the left I struggled to keep the bandit in sight through the top of the bubble canopy. As fighter pilots say, "Lose sight, lose the fight!"

The stall warning buzzer blasted as I shoved the stick further to the left and forward. We rolled and the nose dropped, with

Lake Mathews and its surrounding mountains twisting and turning as we approached the 3000-foot "hard deck" minimum altitude. The buzzer quit, and was replaced by an increasingly loud whistling noise as the plane raced earthward at high speed. I partially leveled my wings to keep the bogey in sight, then yanked back on the joystick. The G-force made me grunt and jammed me down into my seat at over four times my regular weight, making it difficult to keep my head up. The aircraft nose rose and the plane started to buffet as the buzzer screeched again.

Never losing sight of the "enemy," I could see I was closing in, but at a poor deflection angle, so I pulled the plane around and down into a "low yo-yo" to get into a better position behind his beam. I was approaching a firing solution. This guy was dead meat! Again the buzzer sounded as I pulled up (more Gs!) and let the bogey drift down dead center into my 100-mil gunsight. Tracking him in my sight as he turned, I squeezed the joystick trigger. A steady tone, a red panel light, and smoke streaming from the bogey confirmed my kill! "Yee-haw, Jester is dead! Knock it off, knock it off."

This whole sequence took 55 seconds, and was the first of my six dogfights that afternoon with Air Combat U.S.A. I wish I could say I nailed the other guy all six times, but it was three and three. The embarrassing part is that my adversary had never flown as a pilot before!

Air Combat U.S.A is not a simulator—you actually fly, even if you don't have a pilot's license. The planes are two-seat Italian SIAI-Marchetti SF-260W "Warriors," with the "guest pilot" in the left-side pilot's seat and an instructor safety pilot in the right-side seat. The SF-260W is a light attack and tactical support aircraft used by over twenty countries, including several in NATO. It uses a 260-horsepower Lycoming piston engine, and has a 27-foot wingspan with a top speed over 200 mph. It is fully acrobatic and built to handle over 6 Gs.

I was reliving my Navy fighter pilot days of the early 1950s when I flew F4U-5 Corsairs in Fighter Squadron Fourteen (VF-14 "Tophatters.") The SF-260W is a nimble machine compared to the 2000-plus horsepower Corsair. You hardly need to touch the rudder, even in formation, and the stick forces are relatively light compared to a Corsair. Also, most importantly, although the stall buzzer screeches a lot, the plane mushes instead of stalling, and has no spinning tendency. The Corsair, on the other hand, could easily stall and spin, with difficult recovery.

But I found the Gs were more oppressive than in the Corsair, since we wore no G-suits in the SF-260W. This is tough if you haven't pulled any Gs in almost forty years, and you're 64 years old!

Air Combat U.S.A., based near Los Angeles, has put over 3500 guest pilots through their ACM (Air Combat Maneuvering) course in the last two years, with no accidents or near misses (although a few guest pilots have blacked-out for a minute or so during high-G maneuvers.) Pilots and non-pilots are treated essentially alike, and sometimes non-pilots (who follow instructor-pilot prompting without question) "wax" fighter jocks who are used to doing things their way. A patented electronic tracking system is used to confirm the "kill" and turn on the smoke. Also, the entire flight is recorded in the cockpit, switching between two camera angles and the gunsight, and you get to take the VHS video tape home!

Most guest pilots only go through Phase I training, one-versus-one ($400 for four dogfights, $500 for six.) This includes fitting you with a freshly-laundered flight suit, a hard-hat helmet and a parachute. Then you have a one-hour briefing from one of the instructors, most of whom are former military combat pilots. The briefing includes gunsight tracking, high and low yo-yos, pursuit angles, lag rolls, flat and rolling scissors, and energy management. Three other ACM advanced phases are also available, with one flight for each.

The instructors take off and land, but on the way to and from the "combat" area you get to fly formation—something most civilian pilots never have an opportunity to do. Each flight runs about an hour, including practice, instructor demonstrations and dogfights. Then there's a debriefing where both cockpit video tapes are shown together, on two monitors, with a discussion of who did what to whom!

This is great stuff. While you may not have the horses of the Mustangs, Thunderbolts, or Corsairs and Hellcats of WWII, you have a lot more than the Red Baron ever had—and the twisting, rolling, yanking, banking and Gs are all real!

Air Combat U.S.A., Inc., P.O. Box 2726, Fullerton Airport, CA 92633, telephone (800)522-7590.

I can still get into and zip-up my original flight jacket, almost 50 years after it was issued to me!

Six dogfights in an SF-260 Marchetti, and I came down with a torn retina! My last aerobatic flight. Only gliders, home-builts, ultra-lights, and light planes since then . . .

APPENDIX A

ACCIDENT REPORTS #5- #23

12 October, 1950 to 18 February 1952

(Note: Accident Reports #1 and #2 follow Chapter1. Accident Reports #3 and #4 follow Chapter 13.)

The Accident Reports shown in this Appendix are based on photocopies of microfilm from Navy records, purchased from researcher Craig Fuller, 566 March Ave., Healdsburg, CA 95448 (Phone 707-431-0824, email aair@juno.com).

The originals were typed on pre-printed fill-in forms. Unfortunately, the quality of the photocopies was so poor that they were not practical to reproduce here, so I've converted them to a text format.

Some portions of the originals were difficult to read. Also, many abbreviations were used, and the punctuation and grammar were not in the best literary traditions. I've expanded most abbreviations and done some minor editing. However, I believe my "text conversions" are accurate in depicting the content of each report.

It should be noted that ALL of these accidents, plus my two described in Chapter 13 (a total of 21 accidents in 16 months—including two fatalities!) were in F4U-5s in VF-14 while I was in the squadron. I knew all the pilots.

This may give you an idea of the "trial and terror" associated with flying the F4U-5 Corsair from straight-deck "jeep" CVE and "light" CVL aircraft carriers.

Fred "Crash" Blechman

ACCIDENT REPORT #5

Date: 12 October 1950, 21:10

Pilot: ENS Charles P. Moore USNR

Organization: VF-14, NAS JAX, CCVG-1, ComFairJax, ComAirLant
Aircraft: F4U-5 #121868
Purpose: Flight cross country
Hrs.last 3 months: 145.0; Total hours: 721.3
Location: NAS Jacksonville
Weather: Contact
Injuries: None

SPECIFIC ERRORS:
The installation of a neoprene seal in the banjo fitting is not in accordance with the R&M Manual. There is no record in this command of the person or persons who made this improper installation.

CHECKOFF ITEMS:
A 60-degree crosswind from the left caused the plane to start a turn to the left. Soft sand just off the runway caused the plane to nose up.

ANALYSIS:
Pilot had just returned on a cross country flight from NAS

Miami. Normal approach and landing was made, but during rollout the plane began a gradual turn to the left. Pilot used corrective measures with rudder and brake. At that time complete failure of right brake was evidenced. Plane continued to the left and ran off runway into a soft shoulder, causing aircraft to nose up.

Investigation revealed the use of a neoprene seal between the inside crush washer and the brake system of the starboard wheel. According to the R&M Manual a crush washer should be used in this installation.

The brake system was filled and checked. When pressure was applied to the line, a leak was noticed on the inside and outside of the banjo fitting. Although a preflight pilot inspection was held on the plane prior to takeoff from Miami, it was dark and it is doubtful whether the leak would have been noticed, as it was hard to detect in daylight without pressure on the brake system. It is believed that with the constant pressure on the brake system in taxiing out for takeoff and the brakes being held for engine runup, that all fluid was dissipated by the time the takeoff was effected. Consequently, when the pilot tried to use his brakes on the landing rollout, he found he had no pressure in his starboard brake system and this allowed the plane to weathercock into the wind, roll off the runway, and nose up in the soft shoulder.

MATERIAL FACTORS:

Neoprene seal was found on the inside of the crush washer in the banjo fitting. This neoprene seal was deteriorated to such an extent that brake fluid was allowed to escape. There was also a leaky brake seal in right wheel.

There is no provision in R&M Manual for the installation of a neoprene seal in the brake system of this type aircraft.

LOCAL RECOMMENDATIONS:

It is recommended that all personnel concerned be cognizant of the fact that use of a neoprene seal is not an authorized installation in this brake system.

REMARKS:
Damage: Starboard wheel.

ACCIDENT REPORT #6

Date: 23 October 1950, 12:25

Pilot: LTJG Albert E. Hansen USN

Organization: VF-14, NAS, CCVG-1, ComFairJax, ComAirLant, CNO
Aircraft: F4U-5 #121838
Purpose: Hurricane evacuation
Hrs.last 3 months: 45.9; Total hours: 349.2
Location: NAS Jacksonville
Weather: Contact
Injuries: None

SPECIFIC ERRORS:
Pilot allowed his plane to close interval between himself and plane ahead, necessitating use of brakes to avoid collision. In doing so, he applied excessive amount of brake, causing aircraft to nose up.

ANALYSIS:
Pilot had just returned from Atlanta on hurricane evacuation return flight. The flight of 8 made normal breakup and commenced landing procedure. Hansen was #8 in the flight and had made normal landing. During the latter part of his rollout he allowed the interval between his plane and the plane ahead to close. In the effort to prevent collision he

applied both brakes and nosed up. The aircraft returned immediately to a 3-point attitude.

This accident was caused by multiple of errors. The #6 plane was rolling very slow on last quarter of runway, allowing #7 to overtake him. #7 allowed his plane to roll abreast and slightly ahead of #6. LTJG Hansen was dressing off the #7 plane and as #6 plane's wing came into his vision he applied both brakes excessively, causing his aircraft to nose up.

Witness #1 saw Hansen with his head down in cockpit, which explains the reason he was startled when he looked up and applied brakes in such a positive manner.

This accident could have been avoided had all concerned handled their planes in proper manner prescribed by squadron doctrine. However, this does not relieve Hansen of the responsibility for damage to this aircraft.

SPECIAL EQUIPMENT:
Shoulder harness effective.

LOCAL RECOMMENDATIONS:
Reemphasis is again being placed upon complete cockpit familiarization to alleviate the necessity for visually locating the various controls.

GENERAL RECOMMENDATIONS:
In addition to the above statement, this accident indicates the necessity to keep all pilots educated in proper procedures relating to group and squadron operations (i.e, landing intervals, clearing runway, etc.).

COMMANDING OFFICER:
Concur with findings and recommendations of Board. Disciplinary action has been taken. This accident is a direct reflection of an old adage, " You can take a horse to water,

but you can't make him drink." This squadron utilizes a written outline for standard preflight briefing checkoff list that covers every phase of plane handling from taxiing out to takeoff, to cutting engines in chocks. Hansen has been well exposed to current flight procedures established within this squadron, and it can only be considered that it was his own personal errors that led to this accident. I hope and believe that this minor accident will serve as a practical education to our junior pilots, and emphasize to them that correct procedures and established doctrines and briefings be closely followed in order to eliminate these pilot-error accidents.

REMARKS:
Damage: Prop blades.

ACCIDENT REPORT #7

Date: 14 November 1950, 13:15

Pilot: ENS Merle A. Rice USNR

Organization: VF-14, NAS JAX, CCVG-1, ComFairJax, ComAirLant, CNO
Aircraft: F4U-5 #121793
Purpose: Instrument safety pilot
Hrs.last 3 months: 124.7; Total hours: 900.2
Location: 3 miles east of NAS Lee Field, level flight 2000 feet.
Weather: Contact
Injuries: None

SPECIFIC ERRORS:
Bird should have given way to larger less maneuverable flying machine. He was flying AGAINST the traffic in Ground Controlled Approach (GCA) pattern.

ANALYSIS:
This accident occurred while Rice, on routine type GCA training flight, was flying F4U-5 #121793, and acting as Safety Pilot for another pilot conducting type GCA approaches at Lee Field, Jacksonville. Rice was flying at 2000 feet approximately 500 yards astern and to the right of the plane he was observing. Rice was maintaining a lookout for

distant obstacles relative to plane in area he was guarding and was not intent on his immediate surroundings. Rice was looking to port side, when from the corner of his eye he noticed an object in his flight path. He did not have sufficient time to take evasive action. The object (probably a buzzard) struck the leading edge of Rice's starboard wing one foot outboard of the guns. The bird slid over the top of the wing and was lost from view.

Rice regained control of his plane and ascertained that the plane handled normally in all conditions, and continued the flight without further incident. Later inspection revealed that there was a dent in the leading edge of the starboard wing of 2-inches in depth by 8-inches square, where 12 rivets were torn loose from the outer skin.

In an accident of this nature, which might be called "an act of God," had Rice seen the bird far enough ahead he may have had time to maneuver his aircraft and miss it. But at the speeds at which present day aircraft travel, and the minute size of the object hit, this supposition is highly unlikely.

LOCAL RECOMMENDATIONS:

GCA operators be instructed to warn pilots when birds appear on GCA radar screen.

ORDERS:

Bird was flying in control zone opposite to normal flow of traffic, and without positive control or clearance. Ref: CAR 60.18. Bird did not yield right of way to larger less maneuverable aircraft. Ref: CAR 60.14.

COMMANDING OFFICER:

Although these recommendations are submitted in light vein of thought, you can be well assured that accident is not accepted lightly. Pilots of this squadron follow the practice of notifying tower when concentrations of birds are observed in a particular area. It is my recommendation that other squadrons observe this to keep pilots forewarned of this danger.

REMARKS:
Damage: Right wing.

ACCIDENT REPORT #8

Date: 12 December 1950, 15:00

Pilot: ENS Richard S. Kapp, USN

Organization: VF-14, NAS JAX, CCVG-1, ComFairJax, ComAirLant, CNO
Aircraft: F4U-5 #122010
Purpose: Cross country navigational training
Hrs.last 3 months: 116.8; Total hours: 483.5
Location: Runway 19, McCalla Field, Guantanamo Bay, Cuba
Weather: Contact
Injuries: Burns

ACCIDENT:
Pilot was making normal takeoff from McCalla Field for return cross-country navigational flight to Jacksonville. Prior to beginning his initial takeoff run, he had made the necessary pre-flight and engine checks. The tabs were set properly for normal takeoff. Plane started normal run down port side of Runway 19 for approximately 200 feet where it swerved severely to port assuming heading about 40-degrees off initial course. According to pilot's report, he made the necessary corrections of full right rudder and intermittent use of right brake. Also, according to the pilot's statement and one of witnesses' statement, that in order to miss

parked fire trucks and planes, he kept full throttle, right rudder and right brake. The plane came back to original takeoff heading, but was in left skid with tail high, and prop cutting small trenches into ground for approximately 175 feet before hitting a small ditch which collapsed the port landing gear, and nose up followed. Fire from full belly tank broke out immediately and pilot dove head first over port side of cockpit.

It is the opinion of this board that improper takeoff technique was used by this pilot for the F4U-5 type aircraft. Evidently this pilot had made it a practice of forcing tail of plane up with forward stick pressure in order to see down runway. Thus, when plane started running forward with locked tail wheel off deck, there was no positive directional control until aircraft built up enough forward speed for rudder to take effect. It was before this speed was reached that plane got out of control.

As a result of this accident, this squadron has again already briefed all pilots in proper takeoff technique, i.e., allow tail of plane to come up slowly and naturally as plane builds up enough speed for positive rudder control.

COMMANDING OFFICER:

All pilots reporting to this squadron are thoroughly briefed on proper takeoff technique prior to any familiarization flights in F4U-5. The practice of forcing tail up on initial takeoff run is not condoned as you lose all directional control until enough speed is attained for rudder control. All pilots were again cautioned against this practice and it will be made subject for squadron briefings periodically.

REMARKS:

Damage: Strike.

ACCIDENT REPORT #9

Date: 30 January 1951, 10:57

Pilot: ENS Willam L. Cleland USN

Organization: VF-14, USS Wright, CCD4, Com6th Flt, ComFairJax, CAL, CNO
Aircraft: F4U-5 #121947
Purpose: Carrier requalification
Hrs.last 3 months: 68.9; Total hours: 490.1
Location: USS Wright (CVL-49)
Weather: Contact
Injuries: None

ACCIDENT:
Pilot was making carrier approach to USS Wright for carrier landing. It was third landing of requalification flight during which he was to make six landings. The pilot approached ramp for cut at proper altitude and speed. After receiving the cut, pilot, in attempting his landing, flared out too high and floated down the deck, engaging #9 wire. During rollout after engaging wire pilot crashed into barrier.

CONCLUSION:
Both pilot's misjudgment of the relative wind and his flare out technique contributed to this accident. All pilots in this squadron are briefed before and after each flight regarding

carrier takeoff and landing techniques. An experienced pilot observes all takeoffs and landings and comments on the individual techniques displayed.

COMMANDING OFFICER:
Concur.

REMARKS:
Damage: Propeller, speed ring, port wing and wheel pants.

ACCIDENT REPORT #10

Date: 30 January 1951, 13:39

Pilot: LTJG Alfred G. Wellons USN

Organization: VF-14, USS Wright, CCD4, Com6thFlt, ComFairJax, CAL, CNO
Aircraft: F4U-5 #121803
Purpose: Carrier requalification
Hrs.last 3 months: 61.2; Total hours: 360.2
Location: USS Wright (CVL-49)
Weather: Contact
Injuries: None

ACCIDENT:
Pilot was making carrier approach aboard the USS Wright for carrier landing. It was third landing of requalification hop in which he was to make six landings. The tabs were set for landing with full right rudder tab, neutral aileron tab, and 7-degree nose-up elevator tab. The landing checkoff list had been completed and pilot was carrying full rpm, full rich mixture, shoulder harness locked, and wheels, flaps, and hook were in full down position.
Pilot approached ramp for cut in normal attitude and speed, but a little high.
After receiving cut, right wing dropped and although he flew the plane to the deck in a 3-point attitude early enough

to engage #3 or #4 wire, he either had hook bounce or bounced the aircraft enough to miss #4 wire and engaged #5 wire. By the time he engaged #5 wire his momentum to the right carried him into the #3 stack of the carrier.

CONCLUSIONS AND RECOMMENDATIONS:
It is the opinion of this board that possible stack wash and slightly wheels landing were contributing factors to this accident. It is also felt that if corrective action by pilot to get back to center of deck had been started sooner, he would have landed more in the center of the deck, thus avoiding crash into stack when aircraft engaged #5 wire.
All pilots in this squadron have been thoroughly briefed about fact that stack wash may be encountered during carrier passes and also the importance of observing conditions of relative winds and velocities when making carrier landings.

COMMANDING OFFICER:
It is my opinion that this accident is a direct result of pilot error but aided by several contribution circumstances. It is considered possible that this accident may have encountered stack wash at cut as wind was very light and variable throughout the day. The ship was making its own wind and was "chasing wind" through more than a 90-degree sector in an effort to keep the surface wind off the port bow. However, I observed this accident and did not notice stack wash over the deck at this time.
It is my practice to observe every fighter landing from bridge aft or Primary Fly or from LSO platform. I have observed that F4U-5s have tendency for right wing to drop following cut unless flown all the way to the deck. On this particular afternoon, two major damage accidents and one strike damage accident occurred as direct result of right wing drop and pilot's inability, or improper technique, in effecting a

recovery. There were numerous landings when right wing drop occurred, but were either caught early enough to effect proper recovery, or, as on several occasions, the aircraft landed on starboard wheel and hook immediately engaged the first available wire, and starboard moment set by right wing drop was insufficient to carry aircraft into catwalk or stack.

It is recommended that when operating F4U-5 type aircraft from CVLs, pilots should be cut in such position as to allow them to engage the #2 or #3 cross-deck pendant, and no later pendant than the #4. This would allow a greater safety margin to planes that experience wing drop and hook skip.

REMARKS:

Damage: Plane considered to be a strike.

ACCIDENT REPORT #11

Date: 30 January 1951, 15:00

Pilot: ENS Charles P. Moore USNR

Organization: VF-14, USS Wright, CCD4, Com6thFlt, ComFairJax, CAL, CNO
Aircraft: F4U-5 #121793
Purpose: Combat air patrol
Hrs.last 3 months: 67.8; Total hours: 811.4
Location: USS Wright (CVL-49)
Weather: Contact
Injuries: None

ACCIDENT:
Pilot was making carrier approach aboard USS Wright for carrier landing following combat air patrol hop of 1.9 hours duration. His plane was trimmed properly and landing checkoff list had been completed with flaps, wheels, and hook in full down position. Pilot approached ramp for cut at normal altitude and speed on roger pass. After receiving cut his right wing dropped sharply. Pilot applied corrective action to port, and before he could again level off he struck deck, shearing port landing gear as he engaged #4 wire.

ANALYSIS:
Stack wash may have been encountered causing right wing to drop at ramp. Although pilot applied corrective action to

port side in order to line up with center of deck, he either overcontrolled or stalled out at this time and hit the deck in a skid, causing port landing gear to shear off with subsequent damage. Proper use of safety belt and shoulder harness prevented injury to pilot.

CONCLUSION:
In conclusion, the board feels stack wash of flight deck burble caused right wing to drop at the ramp. This fact, plus inability of pilot to land straight ahead in 3-point attitude after taking corrective action, caused the accident. An experienced senior officer of the squadron observes all carrier work and procedures. All pilots in this squadron are thoroughly briefed prior to and after all carrier landings.

COMMANDING OFFICER:
Forwarded. It is my opinion that this accident is direct result of pilot error, but aided by several contributing circumstances. It is considered possible that this aircraft may have encountered stack wash at cut, as wind was very light and variable throughout the day.
The ship was making its own wind and was "chasing wind" through more than a 90-degree sector in an effort to keep the surface wind off the port bow. However, I observed this accident and did not notice stack wash over the deck at this time.
It is my practice to observe every fighter landing from bridge aft or Primary Fly or from LSO platform. I have observed that F4U-5s have tendency for right wing to drop following cut unless flown all the way to the deck. On this particular afternoon, two major damage accidents and one strike damage accident occurred as direct result of right wing drop and pilot's inability, or improper technique, in effecting a recovery. There were numerous landings when right wing drop occurred, but were either caught early enough to ef-

fect proper recovery, or, as on several occasions, the aircraft landed on starboard wheel and hook immediately engaged the first available wire, and starboard moment set by right wing drop was insufficient to carry aircraft into catwalk or stack.

It is recommended that when operating F4U-5 type aircraft from CVLs, pilots should be cut in such position as to allow them to engage the #2 or #3 cross-deck pendant, and no later pendant than the #4. This would allow a greater safety margin to planes that experience wing drop and hook skip.

REMARKS:
Damage: Strike.

ACCIDENT REPORT #12

Date: 11 June 1951, 14:35

Pilot: LTJG James B. Morin USN

Organization: VF-14, USS Siboney, ComFairJax, ComAirLant, CNO
Aircraft: F4U-5 #121838
Purpose: Tactics
Hrs.last 3 months: 112.9; Total hours: 1140
Location: USS Siboney (CVE-112)
Weather: Contact
Injuries: None

ACCIDENT:
Pilot was making carrier approach aboard USS Siboney for carrier landing following tactics hop of 1.9 hours duration. His plane was trimmed properly and landing checkoff list had been completed with flaps, wheels, and hook in full down position. Pilot approached ramp with proper altitude and speed. Just as pilot received cut, ship pitched with stern down and bow up. By the time pilot nosed over in order to reach the deck, the ship's deck started up and pilot crashed. Both landing gear collapsed.

ANALYSIS:
Contributing factors were: (1) Ship was pitching due to

gentle swells making it difficult for pilot and LSO to judge correctly when to cut and how to land; (2) Pilot continued to hold right rudder after cut, thereby causing plane to fly right with possible twisting pressure on landing gear upon impact; (3) Pilot dove for deck due to its pitching and because he felt high. He did not flare out in time due to reverse motion of the deck.

CONCLUSIONS:
The board feels the pitching deck condition together with diving for the deck were contributors factors to this accident. It is evidenced how small a margin of safety was present at time of operations following Captain's decision to send the remaining Corsairs to the beach after this accident. All landing and takeoff techniques are observed by experienced aviators in this squadron, and thorough briefing prior to and after all launches is conducted.

COMMANDING OFFICER:
It is always easier to second-guess an accident, but these remarks picture my impression of the accident at the time it occurred. Approaching the groove, Morin was high and allowed his plane to drift right. Just before the cut the ship's stern started down, placing Morin very high. On the cut, he dived for the deck, landing in very flat attitude, on main wheels, just short of #3 wire, still slightly right.
LSO point out that ship was pitching through an arc of 20-60 feet. He further states that Morin did not immediately answer cut by nosing over. In view of these remarks and from my own observations, I attribute this accident equally to the LSO and Morin. Morin was worked too close to the safety margin. With a pitching deck removing much of the allowable safety margin, LSO should have been more exacting as to Morin's position at the cut, in order to eliminate as much pilot error as possible. Following cut, had Morin nosed

over promptly and been able to assume a better attitude prior to contact with deck, aircraft damage may have been materially lessened.

I recommend that LSOs during carrier qualification or refresher landings insist that their pilots be in near perfect condition at the cut or else be waved off. This will give pilot benefit of greater margin of error (i.e., did not relax right rudder, did not immediately nose over for deck) and still make safe landing.

REMARKS:

Damage: Strike.

ACCIDENT REPORT #13

Date: 12 June 1951, 15:37

Pilot: ENS Berkeley W. Hall USNR

Organization: VF-14, USS Siboney, ComFairJax, ComAirLant, CNO
Aircraft: F4U-5 #121888
Purpose: CIC Tactics
Hrs.last 3 months: 106.4; Total hours: 656.1
Location: USS Siboney (CVE-112)
Weather: Contact
Injuries: None

ACCIDENT:
Pilot was making an approach aboard the USS Siboney for carrier landing following CIC and Tactics hop of 2.0 hours duration. His plane was trimmed properly and landing checkoff list had been completed with flaps, wheels, and hook in full down position. Pilot made normal carrier pass at proper altitude and speed. After cut the pilot endeavored to land, but after engaging #8 cross-deck pendant crashed into #2, #3, and #4 barriers.

ANALYSIS:
Upon analyzing the accident, the pitching deck and low relative wind were definitely contributing factors causing

the accident. Pilot could have reached deck with added risk of "diving" for deck with possible structural damage as result of a hard landing.

CONCLUSIONS AND RECOMMENDATIONS:
It is the opinion of this board that pilot was not completely at fault. As the Aerological Officer's statement shows, only 5 knots of surface wind were present at the time of the accident. LSO has stated he only had an indicated relative wind of 21 knots over the deck. Although 25 knots of relative wind are considered ample to operate Corsairs from carrier, this board feels consideration should be made with respect to the difference in weight and airspeed at the cut between the F4U-4 and F4U-5. This board, therefore, recommends relative wind of at least 27 knots for the F4U-5 when operating from CVE or CVL type carrier.
The slight wind condition and fact that ship pitched forward with bow down and stern up gave Hall the unusual position of flying down hill trying to catch the deck. Captain of the ship sent remaining Corsairs that were airborne (flight of 10) to beach. All pilots in this squadron are thoroughly briefed prior to and after all launches, and in addition are observed by an experienced aviator during all phases of their carrier work with comments on improper techniques, and instructions are given when needed.

COMMANDING OFFICER:
Concur

REMARKS:
Damage: Engine, propeller, landing gear fairings.

ACCIDENT REPORT #14

Date: 11 July 1951, 12:45

Pilot: LTJG Albert E. Hansen USN

Organization: VF-14, NAAS Cecil Field, Jacksonville, ComFairJax, ComAirLant, CNO
Aircraft: F4U-5 #121861
Purpose: Bombing and Tactics
Hrs.last 3 months: 149.4; Total hours: 633.2
Location: NAAS Cecil Field
Weather: Contact
Injuries: None

ACCIDENT:
Hansen, flying side #415, was leader of 2nd division of a 7-plane. Flight was cleared to NAAS Cecil Field Tower to break up and land on right side of Runway 5. After breakup #1 plane (side #409) landed on right side of runway and turned off to right and held at intersection of Runway 27. #2 plane landed on left side of runway and turned left to return to line. #3 plane landed on right and turned right to hold at intersection of Runway 27. #4 plane landed and followed #2 plane to left.
As #5, Hansen (flying #415) landed on right side, tower cleared #1 plane (#409) across Duty Runway 5. As plane was crossing, Hansen cleared behind and began to cross to

left side of runway for his turn off, in such a way that plane in front of him was hidden by his nose. As he was completing his turn off duty runway he was very close to #1 plane and was told by #1 plane to "Hold it 415." Pilot applied brakes and nosed up.

ANALYSIS:
Upon analyzing this accident, pilot taxiing across duty runway, and tower giving clearance when other planes were landing, were contributing factors. Also, pilot cleared himself across duty runway instead of waiting for tower clearance. Pilot did not have plane well enough under control to stop on a moment's notice.

COMMANDING OFFICER:
It is my confirmed opinion that a taxi accident is a result of errors of pilot immediately involved. That is true in this case, although errors of other personnel made this accident more probable. It is doctrine in this squadron that pilots land their aircraft on the first third of the runway, complete landing roll-out within 2nd third, and have aircraft under positive taxi control for turn off runway. Duty runway in this instance, Runway 5, does not have a taxi strip at the end of the runway, consequently tower-approved turn offs were executed at intersection of Runways 5 and 27. This, in effect, a closer turn off interval.
Tower made first of a series of errors when clearance was given to side #409 to cross duty runway from his parked position at head Runway 27, while other planes in flight were still landing. Pilot of side #409 made next error when he elected to take tower approval to taxi across duty runway directly ahead of oncoming planes. Hansen then committed next error by turning across runway from starboard to port without tower approval. I believe he missed seeing #409 cross ahead, as he was checking behind for enough

interval to permit a left turn off. Hansen had plane under proper control for turning off and would have been in satisfactory position had he observed #409. Hansen reacted too violently on brakes, physically causing nose-up. Hansen has been properly admonished, and all pilots have again been briefed on proper procedure and safety.

REMARKS:
Damage: Propeller

ACCIDENT REPORT #15

Date: 20 July 1951, 12:16

Pilot: LTJG Donald S. Ross USN

Organization: VF-14, NAAS Cecil Field, Jacksonville, ComFairJax, ComAirLant, CNO
Aircraft: F4U-5 #121843
Purpose: Fighter escort
Hrs.last 3 months: 90.0; Total hours: 466.1
Location: NAAS Cecil Field
Weather: Contact
Injuries: None

ACCIDENT:
Ross was in last plane of an 11-plane combat escort hop which had just returned from a routine training mission. Immediately after breaking to downwind leg, Ross placed landing gear lever in down position. While going over check-off list he noticed that indicator showed gear to be in up position. After Tower was notified, Ross climbed to 1000 feet to check hydraulic system, which was found to be in proper working condition. Thereupon Ross reached under left console to check rear of gear lever. Lever itself was found to be broken loose from rest of actuating linkage.
At this, Ross realized that it would be impossible to lower gear, as emergency procedure is dependent upon lever to

trip it into action. Tower was notified that a wheels-up landing would have to be effected. Traffic was cleared and Ross went through normal landing procedure except for a slightly longer straightaway. Just prior to touchdown, mixture control was placed in idle cut-off position and all switches were placed in off position. After touchdown aircraft skidded to a stop in approximately 300 feet. Pilot cleared aircraft immediately.

ANALYSIS:
Opinion of board that accident was caused by improper installation is backed up by fact that nut from bolt which connects pilot's landing gear control lever to lever assembly was found missing. It is believed that since aircraft's last overhaul, nut backed off due to improper installation. This would be entirely possible as nut should have been of self-locking type. It is conceivable, due to haste, that wrong type of nut was placed on bolt.

COMMANDING OFFICER:
In addition, all of remaining planes were immediately inspected to determine condition of specified bolt, and all were found to be in excellent condition. This particular part has been added to 120-hours check for inspection in order to safeguard against any similar accident. Chance-Vought Representative has been informed of this accident and has recommended a manual release be incorporated so that lever assembly may be actuated by hand should any similar failures occur to pilot's landing gear control lever.

REMARKS:
Damage: Four propeller blades. Sudden stoppage of engine. Sway braces and landing flaps.

ACCIDENT REPORT #16

Date: 4 September 1951, 09:45

Pilots: LTJG Charles P. Moore USNR and

LTJG Donald R. Tate USNR

Organization: VF-14, NAAS Cecil Field, ComFairJax, ComAirLant, CNO
Aircraft: F4U-5 # 121831 (Moore); F4U-5 #121989 (Tate)
Purpose: B-29 Intercept
(Moore) Hrs.last 3 months: 156.3; Total hours: 1107
(Tate) Hrs.last 3 months: 103.9; Total hours: 845.5
Location: 12 miles SSW St. Augustine, Florida
Weather: Contact
Injuries: Tate: Fatality; Moore: Bailout, uninjured

ACCIDENT:
Moore departed with 4 F4U-5s from NAAS Cecil Field at about 0830 on 4 September for B-29 Intercept flight. Periodically this squadron has worked with Air Force on Fighter Interception of B-29 Bombers for purpose of training both types in intercept problems. All of the VF-14 pilots were considered qualified to participate in this type of maneuver. LT Bennett was flying as wingman on LTJG Moore. LTJG Riddick was section leader and LTJG Tate was flying #4 position.

Prior to the accident, the flight had effectively completed three intercepts. It was after the second run on the third intercept that the accident occurred. It appears that during the subsequent rendezvous, aircraft were somewhat strung out and #4 man, Tate, over ran his section and joined on lead man, Moore. While Tate was flying close to, but slightly astern of , Moore, he recognized that he had joined on the wrong aircraft. It is believed by this board that Tate was looking over his left shoulder to determine which aircraft was his section leader.

At this time, Drexel Easy, Ground Controller, vectored the flight to a new intercept with a hard port turn to 60-degrees. Moore turned immediately to the new vector and at that instant the two aircraft came in contact. Immediately after the accident, which occurred at 18,000 feet, Moore successfully bailed out. Apparently Tate made no attempt to leave the wreckage, as evidenced by the fact that his body was still into the pilot's seat.

ANALYSIS:

After completion of interception run, it can be noted that aircraft are strung out considerably. It is conceivable that last aircraft was undecided as to which aircraft was his section leader. In his haste to join up, he joined on the wrong aircraft. Realizing his error, Tate turned his head away from the aircraft he was flying wing on, to locate his section leader. Had Tate kept his eye on the aircraft in front of him, and maneuvered to a safer distance before looking around, this accident would not have occurred. It is recommended by this board that all Fleet squadrons be cognizant of fundamentals of formation flying. Aviators who have many hours in formation flight can have the same type of accident if they are lulled into a sense of security because of their own skill in this particular maneuver.

COMMANDING OFFICER:
I concur with the opinion of the board in that Tate may have erred in joining on the division leader during this tactical maneuver. In attempting to locate his section leader, Tate may have relaxed his proper caution and vigilance. I do not believe that other pilots in this flight were responsible in any way for this accident. I considered the pilots involved in the accident fully qualified to carry out the mission assigned. No disciplinary action is considered necessary and none has been taken.

REMARKS:
Damage: Both aircraft strikes.

ACCIDENT REPORT #17

Date: 4 November 1951, 17:11

Pilot: LTJG Alfred G. Wellons USN

Organization: VF-14, USS Kula Gulf, ComFairJax, ComAirLant, CNO
Aircraft: F4U-5 #121947
Purpose: Combat air patrol
Hrs.last 3 months: 112.1; Total hours: 624.9
Location: USS Kula Gulf (CVE-108)
Weather: Contact
Injuries: None

ACCIDENT:
LTJG Wellons was the number 2 man in a flight of four F4U-5s. Upon completion of a routine combat air patrol flight, the aircraft entered the pattern for a normal carrier landing, and started to climb while entering the groove. The LSO gave the "high" signal and the pilot answered properly. When LTJG Wellons reached the ramp, he was in a good position for a cut, which he received. After taking the "cut" LTJG Wellons was slightly left of center and dropped his right wing to fly back to the center line of the deck. In doing so, he stopped his rate of descent by holding the nose of the aircraft up, allowing the aircraft to proceed up the deck. LTJG Wellons dropped his nose to lose the remaining

few feet of altitude and engaged the number 6 cross-deck pendant, pulling out enough wire to allow contact with the number one barrier.

At this instant all four barriers were in the up position since the preceding aircraft had just cleared the barriers forward. Before declaring a clear deck, three barriers must be in the up position. At the time of the accident, the first three barriers were up, and the fourth was on the way to the up position. There was not sufficient time to drop the first barrier during the landing and still have three barriers in the up position. The interval between the aircraft was normal, but the crowded condition of the deck forward of the barriers prohibited the pilot of the taxiing aircraft to taxi smartly into the spot. Because of the slow relative speed of the aircraft and the fact that the aircraft had engaged the number 6 cross-deck pendant, the damage was minor. Upon contact with the number one barrier, one blade of the prop snapped off, and a few dents and tears were inflicted to the speed ring, necessitating a change of both parts. No other damage was sustained.

It is the opinion of this accident board that LTJG Wellons, while attempting to fly back to the centerline of the deck, allowed the aircraft to proceed further up the deck to assure positive control of the landing. In so doing, the aircraft engaged a late wire, which under normal conditions would not have resulted in a barrier crash. Had the number one barrier been in the down position this accident would not have occurred.

ANALYSIS:

Landing separations aboard a CVE aircraft carrier require a longer landing interval between aircraft because of congestion in the restricted parking area forward of the barriers. LTJG Wellons had a landing interval of 30 seconds, which was sufficient for a normal arrested landing, but not for a crowded deck forward. It is the opinion of this board that LTJG Wellons erred in judgment by holding off while attempting to fly back to the centerline. Also, he was not sufficiently left of center to warrant this correction.

CONCLUSION:
This board recommends that all flight squadrons remain cognizant of the points brought out in this aircraft accident report. (1) The technique involved during the critical period between the "cut" and the arrested landing should be periodically reviewed by all squadrons for the respective type aircraft. (2) While operating aboard a CVE a shorter parking area forward of the barriers. This squadron recommends forty seconds as a safe and expeditious landing interval.

REMARKS:
Damage: Propeller, speed ring.

ACCIDENT REPORT #18

Date: 6 November 1951, 11:39

Pilot: LTJG Charles P. Moore USNR

Organization: VF-14, USS Kula Gulf, ComFairJax, ComAirLant, CNO
Aircraft: F4U-5 #121949
Purpose: Combat air patrol
Hrs.last 3 months: 132.8; Total hours: 1199.7
Location: USS Kula Gulf (CVE-108)
Weather: Contact
Injuries: None

ACCIDENT:
LTJG Moore was the section leader of a four aircraft division returning from a combat air patrol flight. The weather was not good as there were many rain squalls around the ship. At the time of the "Prep Charlie," the ship was in a clear area and heading toward a rain squall. During the landing operation there was light to moderate rain over the flight deck and at the time of Moore's landing the relative wind was variable from 25 to 35 knots across the deck, and 20-degrees from either side of the bow.
LTJG Moore's approach was normal but a little fast, necessitating a fast signal from the LSO, which was answered. Upon reaching the ramp the aircraft began to climb at which

time the LSO gave LTJG Moore a "cut." Taking the cut in a little higher position than usual, LTJG Moore decidedly dived for the deck, hitting hard and slightly wheels first, which caused him to bounce back into the air without engaging a cross-deck pendant. The bounce occurred between number 2 and 4 cross-deck pendants and the aircraft remained airborne until passing number 7 pendant.

At this time the pilot realized he was approaching the barriers and nosed over to assure contact with them. The aircraft engaged number 2, 3, and 4 barriers and nosed up because it had not engaged a cross-deck pendant. The aircraft fell back to a three-point attitude after its forward motion had stopped.

ANALYSIS:

The weather and wind conditions may have been contributing factors as to the damage sustained, but it is the opinion of this accident board that pilot error was the primary cause. LTJG Moore was in a fair position for the "cut" and could have made a normal landing. The pilot erred in technique by making a decided dive for the deck to lose the slightly excessive altitude which could have been overcome by the normal flare-out technique. The aircraft hit the deck in a slightly wheels first attitude between number 3 and 4 cross pendants.

The initial contact with the deck was of sufficient force to cause an extremely high bounce which prevented contact with a cross-deck pendant. It was at this point the pilot realized that an arrested landing was impossible and nosed over into the barriers. The use of the shoulder harness, safety belt, and E-3 protective helmet was responsible for the lack of personal injuries in this accident.

CONCLUSION:

In the opinion of this accident board, diving for the deck was the primary factor causing this accident. Again, this accident board recommends that all flight squadron commanders emphasize the importance of flying the aircraft from the "cut' until a normal full-stop arrested landing has been effected.

COMMANDING OFFICER:
Concurs.

REMARKS:
Damage: Sudden stoppage of engine necessitated an engine and prop change. Landing gear, port outer wing panel.

ACCIDENT REPORT #19

Date: 6 November 1951, 12:11

Pilot: ENS James L.Ellis USNR

Organization: VF-14, USS Kula Gulf, CAG-1, ComFairJax, ComAirLant, CNO
Aircraft: F4U-5 #121561
Purpose: Combat air patrol
Hrs.last 3 months: 80.1 ; Total hours: 606.9
Location: USS Kula Gulf (CVE-108)
Weather: Contact
Injuries: None

ACCIDENT:
ENS Ellis was number 4 man in a flight of four F4U-5s returning from a combat air patrol flight. Landing was delayed because of a barrier crash on deck. When the "Charlie" signal was given ENS Ellis started an approach to the ship. He received a waveoff on the first pass for being low and long in the groove. At the waveoff one of the LSOs dived into the net and the screen had to be dropped to prevent the aircraft from carrying it away.
The next pass was normal until the aircraft reached the groove where it began to settle. Just prior to this the LSO gave a fast signal which was answered. The LSO started giving a "come on" and "low" signal when the aircraft began

to settle. At this point the pilot did not seem to respond to the LSO signals. The landing gear struck the ramp round-down followed by the tail section. At the impact the tail section bounced up into the air with the landing gear remaining on the deck causing the prop to strike the flight deck. The prop continued to hit the deck as the aircraft proceeded forward with the tail section high. The aircraft continued in this attitude until it engaged number 2, 3, and 4 barriers. The tail swung to the starboard and upward, striking the after part of the open bridge. When at a complete stop the aircraft remained in a nosed up attitude, pointing thirty degrees to the port of the bow.

ANALYSIS:

ENS Ellis made a normal approach until he started to approach the groove. His speed was a little on the fast side necessitating a fast signal which he answered. At this point his aircraft began to settle. Because of the near proximity of the aircraft to the ramp the LSO was unable to give a waveoff. The LSO gave one "comeon" signal followed by violent "lows" in an attempt to get the aircraft up and over the ramp. The aircraft struck the ramp while the LSO was still giving the "low" signal.

It is the opinion of this accident board that the pilot erred in technique in answering signals, in that he eased off too much power on the fast signal, and did not respond properly to the following "comeon" and "low" signals, causing the aircraft to settle into the ramp. The use of shoulder harness, safety belt, and E-3 protective helmet was responsible for the lack of injuries in this accident.

The primary factor causing this accident was improper handling of the aircraft in response to the LSO signals. The accident board strongly recommends that all flight type squadron commanders continually review the proper technique and procedures in answering the LSO signals, and the importance of doing so immediately. Also, to recognize the "low" and "slow" as danger signals. All pilots should make every effort not to get into such a position close to the "cut" position.

COMMANDING OFFICER:
Concurs. ENS Ellis is considered to have been properly indoctrinated and qualified for carrier operations in the F4U-5.

REMARKS:
Damage: Strike

ACCIDENT REPORT #20

Date: 13 November 1951, 13:10

Pilot: LT Jesse R. Hopkins USNR

Organization: VF-14, USS Kula Gulf, CAG-1, ComFairJax, ComAirLant, CNO
Aircraft: F4U-5 #122164
Purpose: Combat air patrol
Hrs.last 3 months: 157.2 ; Total hours: 2095.4
Location: USS Kula Gulf (CVE-108)
Weather: Contact
Injuries: None

ACCIDENT:
Lt Hopkins was leading a flight of four F4U-5s and had just arrived at base for recovery. His approach to the ship was a little long in the groove. After the cut LT Hopkins made a good landing but allowed the tail to rise shortly after contact with the deck. The initial contact was between number 3 and 4 cross-deck pendants. The hook bounced over number 4 pendant and it was at this point where the tail began to rise. The pilot seemed to nose over slightly to a level attitude with the main landing gear still on the deck. The aircraft continued in this attitude until it reached the barriers (numbers 2, 3, and 4) where it nosed up abruptly and fell back to a three point attitude.

ANALYSIS:

LT Hopkins made a very good landing between number 3 and 4 cross-deck pendants. The tail hook bounced over number 4 pendant, which may have been a contributing factor in this accident. Just after passing number 4 pendant the pilot nosed over slightly and held the aircraft in level attitude as it proceeded up the deck, preventing contact of the hook with the remaining pendants. The pilot believed it too late to engage a pendant and nosed over to assure contact with the barriers. The pilot erred in technique by not holding the aircraft on the deck in a three-point attitude, eliminating the possibility of engaging the remaining cross-deck pendants. The use of shoulder harness, safety belt, and E-3 protective helmet was responsible for the lack of injuries in this accident.

The primary factor causing this accident was improper technique, in that the pilot raised the tail off the deck preventing engagement of the remaining cross-deck pendants. This board recommends that all carrier pilots flying conventional type aircraft be aware of the dangers of raising the tail after the aircraft is on the deck in a three-point attitude. With the tail high there is no possibility of engaging the arresting gear, thus inviting an accident.

COMMANDING OFFICER:

Concurs. All pilots are briefed, immediately following each carrier recovery, regarding individual pilot technique. Those pilots not flying observe launches and recoveries.

REMARKS:

Damage: Sudden stoppage of engine, propeller. Port wing was tangled in the wires of the barriers.

ACCIDENT REPORT #21

Date: 11 January 1952, 15:02

Pilot: LT Emmet Brown USNR

Organization: VF-14, USS Wasp, CAG-1, ComAirLant, NASA
Aircraft: F4U-5 #121980
Purpose: Strike tactics
Hrs.last 3 months:134 ; Total hours: 2137
Location: USS Wasp (CV-18)
Weather: Contact
Injuries: None

ACCIDENT:
LT Brown was the second division leader of a two-division flight returning to the ship after a routine tactics flight. His breakup and approach to the landing were normal, arriving at the cut position in a slight wing down attitude. As the cut was given, Brown nosed over slightly and the aircraft landed in a left wing down and nose down attitude. The port landing gear sheared off, and the aircraft engaged the number 2 cross-deck pendant. Just after engaging the pendant the prop dug into the deck and two blades were snapped off near the tips. The aircraft then skidded eighty feet beyond this point and came to rest with the port external tank supporting the left wing.

ANALYSIS:
Brown made a normal approach to his landing, arriving at the cut position with good altitude and speed but slightly port wing down. As the cut was given he overestimated his altitude above the ramp and consequently set up an abnormal rate of descent. Brown then concentrated on attempting to raise his port wing and inadvertently allowed the aircraft to fly onto the deck in a nose down attitude. It is the opinion of the board that Brown erred in technique by allowing his nose to drop so low that it could not be returned to the three-point attitude before making contact with the deck. Emphasis was placed on the wing low position rather than on both the wing low and nose low attitude. The use of safety belt, shoulder harness, and E-3 helmet insured lack of injuries to pilot.

CONCLUSIONS:
Primary factor causing the accident was the improper technique of allowing the nose to drop too low, thereby creating an excessive rate of descent. Brown states that he attempted to pick up a little extra airspeed for control purposes. This was improper technique as neither the time nor altitude from the cut to the landing permits such late correction. Brown's inability to raise his port wing after the cut could possibly be attributed to stackwash or burble from the stack, for which pilots had received waveoffs. All pilots in this squadron have been reminded of the consequences resulting from allowing the nose to get too low after the cut.

COMMANDING OFFICER:
Concurs.

REMARKS:
Damage: Major overhaul.

ACCIDENT REPORT #22

Date: 2 February 1952, 15:15

Pilot: ENS Richard S. Kapp USN

Organization: VF-14, USS Wasp, CAG-1, ComAirtLant, NAASA
Aircraft: F4U-5 #122158
Purpose: Bombing towed sled
Hrs.last 3 months: 98.8 ; Total hours: 875.2
Location: USS Wasp (CV-18)
Weather: Contact
Injuries: Fatal

ACCIDENT:

ENS Kapp was deck launched from the USS Wasp as the eighth aircraft of a 12 aircraft bombing formation. Each aircraft was loaded with one 500-pound water-filled bomb with which to strike a towed spar at 15:15. At the pre-flight briefing it was stated that the attack would consist of one individual glide bombing run. Due to the small size of the target, and the safety cone involved, all pilots were advised to take a long enough interval to ensure that each pilot picked up his own lead on the target.

The flight leader announced "commencing attack" at which time the first division commenced their dive from 8500 feet and made close but individual bombing runs on the spar as briefed. The second division was close behind with the same type run. ENS Kapp's aircraft was seen to drop its bomb and commence a pullout to a near level attitude at which time the port wing dropped violently until reaching an inverted position and in approximately an 80-degree angle of dive. The aircraft paused momentarily in this attitude, and then commenced another half roll to the left until the wings were in level attitude, but still in a steep dive. The angle of dive was decreased to about 45 degrees at which time ENS Kapp crashed into the sea.

ANALYSIS:

ENS Kapp as fourth man of the second division commenced his dive properly and continued so until the time of bomb drop. At this time he flew his aircraft under and close behind the third aircraft. After his bomb released and during the first part of his recovery, it is believed he encountered slipstream causing his port wing to dip violently until reaching an inverted position. ENS Kapp then continued the roll until the wings were level but in a very steep dive with only about 500 feet of altitude remaining. Kapp then commenced a sharp pull out, but due to his low altitude he could not raise the nose higher than the 45 degree position, and hit the water in this attitude. Upon impact the aircraft disintegrated and sank immediately.

It is the opinion of this board that ENS Kapp erred in allowing his aircraft to get too close and under the aircraft ahead, thus hitting its slipstream. This slipstream decreased the lift on the port wing which, with the increased load factor, resulted in a violent roll. ENS Kapp used proper technique in recovery from the roll, but with insufficient altitude to permit a safe recovery.

CONCLUSION:

The squadron doctrine states that all bombing runs will be completed and the aircraft straight and level by 1000 feet. This board feel that ENS Kapp started to recover at the proper altitude, but that the resulting maneuver did not give him sufficient altitude to recover. All pilots have again been reminded of the dangers of flying close behind and in the slipstream of another aircraft. The possible results of slipstream plus g-forces upon the stalling characteristics of the airfoil have also been discussed.

COMMANDING OFFICER:
Concurs. The preflight briefing is stated to have prescribed individual glide bombing with long enough interval to insure that each pilot picked up his own lead. Accurate and safe bombing against a single towed target dictates against simultaneous or near runs.
Evidence indicates that this accident resulted directly from ENS Kapp following too closely the preceding aircraft. During pull out Kapp encountered the turbulence created by an aircraft closely preceding him, which together with his own high acceleration at pull out resulted in a high speed stall and loss of control at low altitude.
It is recommended that this accident be given wide publicity in appropriate aviation safety bulletins.

REMARKS:
Damage: Strike

ACCIDENT REPORT #23

Date: 18 February 1952, 18:13

Pilots: LCDR Vernon E. Binion USN

and LCDR John C. Kennedy USN
Organization: VF-14, USS Wasp, CAG-1, ComAirLant
Aircraft: F4U-5 #121860 (Binion) and F4U-5 #122162 (Kennedy)
Purpose: Combat air patrol and close air support
(Binion) Hrs.last 3 months:114; Total hours: 1346
(Kennedy) Hrs.last 3 months: 111.7 ; Total hours: 1618.9
Location: USS Wasp (CV-18)
Weather: Contact
Injuries: None

ACCIDENT:
LCDR Binion in #121860 was section leader of a four aircraft combat air patrol flight on 18 February 1952. The flight returned to the ship at 18:05 at which time LCDR Binion made a normal break up and carrier landing. After the cut the aircraft engaged the number 1 cross-deck pendant. LCDR Binion then picked up his hook, raised wing flaps, opened cowl flaps, and added power to taxi up the deck. As the aircraft moved forward he placed the wingfold handle in the fold position and as he reached a point between the forward edge of the number 3 elevator and the number 1 barrier the engine was cut to idle and brakes applied.

The aircraft skidded over the barriers and into #122162, which was in the final parking phase adjacent to the forward edge of the deck-edge elevator. LCDR Binion's prop cut into the tail of #122162 and stopped between the tail and the starboard wing of #122162. Both pilots cut all switches and cleared their aircraft.

ANALYSIS:

LCDR Binion made a normal arrested landing and used proper technique in commencing his taxiing forward. After he initially added power he erred by removing his hand from the throttle control while at high power setting, and allowed his aircraft to gain excessive forward speed. By the time he retarded the throttle control his forward speed was so great that he was unable, even with full brakes, to come to a stop before hitting the aircraft parked forward.

CONCLUSIONS AND RECOMMENDATIONS:

It is the opinion of this board that LCDR Binion erred by allowing his aircraft to gain an excessive taxi speed for the short distance in which he had to stop All pilots in this squadron have been advised to keep their hand on the throttle control until it has been cut to the idle position. LCDR Binion's remarks on how he may have prevented the accident once the forward speed became excessive are very noteworthy. However, it is questionable if the pilot has time to drop his hook or cut his engine in the short interval of time between the instant he realizes that he is fast, and the impending accident.

COMMANDING OFFICER:

Concurs.

REMARKS:

Damage: #121860: Prop necessitating change; #122162: Major overhaul.

APPENDIX B

"Vought F4U Corsair History"

(Internet: www.aviation-history.com/vought/f4u.html)
(With permission from copyright owner Larry P.Dwyer)

The name Corsair has been applied to a succession of United States Navy aircraft through many decades, but none has been more devastating to an enemy than the F4U Corsair, a distinctive 'cranked wing' monoplane fighter.

The Corsair's most unique feature was the "bent" wing, the result of a marriage between the most powerful engine ever installed in a piston-engined fighter at that time, and one of the biggest propellers in the world. The inverted gull wing permitted the short, sturdy undercarriage required for carrier operations, allowed a low-drag, 90 degree wing-fuselage junction, gave the pilot better visibility over the wing and lowered the overall height of the folded wing. An added asset of the gull wing was a planing action during emergency water landings.

Development of the F4U began in 1938, when the US Navy ordered a new carrier-based fighter. The prototype first flew in May 1940, and by February 1943 the aircraft was operational with the US Marine Corps against the Japanese in the Pacific area, flying initially from land bases as the US Navy considered it too fast for carrier operations. This was proved wrong when Corsairs delivered to the Royal Navy were flown from carriers with great success from April 1944.

The Corsair's distinctive sound earned it among the Japanese the nick-name of "Whistling Death", partly because of the engine sound. The sound was caused by the wing-root inlets for engine air. Inside of these inlets were placed the oil coolers which ejected hot air through adjustable doors under the wings just ahead of the spar. However, to the ears of American G.I.s clawing their painful way off the bloody beaches of Okinawa in April 1945, it was the sweetest sound in the world. In the shadow of their "Sweetheart's" cranked wings they found a brief respite from the danger that threatened them from every palm-grove and every scrub-covered ridge; but it is doubtful if any of them realized that the aeroplane which protected them had at one time been officially a "failure".

Armament consisted of six 5 in. Browning M2 machine guns, three in each outer wing panel, normally boresighted to converge at 300 yards. Inboard and intermediate guns carried 400 r.p.g. and outboard guns, limited by wing contours, carried 375 r.p.g. If desired the outboard guns which most affected stability and flutter characteristics, could be removed. The prototype XF4U-1 first flew on 29th May, 1940 with Lyman A. Bullard at thecontrols and its impressive speed of 405 m.p.h. gave the lie to the prevalent theories among Army Air Corps authorities that the future of high-speed fighter design lay in the H-3730 and X-1800 liquid-cooled engine projects. Pratt and Whitney were consequently permitted to cancel these latter projects.

Ease and speed of mass production had not figured largely in the design as war was not considered imminent; thus a U.S. Navy request of 28th November machine, Bureau of Aeronautics Number (BUNO) 02153, flew on June 25,1942 with a top speed of 415 m.p.h., a sea-level climb rate of 3,120 ft./min. and a service ceiling of 37,000 ft. The first carrier trials were carried out on September 25,1942 aboard the U.S.S. Sangamon (CVE-26) by Lt. Sam Porter in BUNO 02159, the seventh production machine. These trials drew attention to a number of problems which prevented the Corsair from going into carrier service with the U.S.N. for some years. The landing gear shock struts were too stiff; and there was a landing "kick" caused by local stall in the crank of the gull wing in the high three-point attitude. Experienced pilots learned to master this but accidents were numerous during training.

The main landing gear legs rotated through 90 degrees as they folded rearward to permit the wheels to lie flat in the wings. The rearward-folding tail-wheel, with the arrester hook attached to the strut, was self-centring, lockable and 360 degree swiveling with a 12 1/2 x 4 1/2 in. pneumatic tire; early models carried an 8 1/2 x 4 in. solid tire.

Very early production Corsairs were equipped with the Brownscope wide-angle rear-view periscope system, which was shortly replaced by a mirror in a small "bubble" in the sliding canopy. The first aircraft to be fitted with the blown hood was BUNO 17456, and the first production aircraft was accepted on August 9,1943. The low cabin line and long nose of the early versions made accurate deflection shooting extremely difficult; the sight line was therefore raised 5 in. and the seat adjust increased to 9 in. Due to the urgency of production requirements the rudder pedals were moved aft but raised only 1/2 in. Pilots thus adopted an almost "standing" posture when the seat was at full height and this, coupled with the fact that theCorsair had no cockpit floorboards gave one the impression of sitting on the edge of a deep pit with a yawning black chasm below. The pilot's posture was satisfactory for long flights but the more nervous felt a constant nagging fear that if they slipped they might just wind up somewhere in the mysterious depths of the fuselage bottom !

On February 13,1943 Marine Fighter Squadron 124 (VMF-124) demonstrated their superiority over the Wildcat by escorting PB4Y-1 Liberators all the way to Bougainville. The following day they saw combat for the first time, and the inexperienced Corsair pilots were badly mauled by some 50 Mitsubishi Zeros. Two Corsairs, two Liberators two P-40s and four P-38s were lost in this "Saint Valentines Day Massacre", but the Corsairs soon gained an ascendancy over the Japanese which they never lost, VMF-124 being subsequently credited with 68 kills against a loss of four aircraft and three pilots. Within six months all Pacific-based Marine fighter squadrons had been re-equipped with the Corsair.

Since Navy Fighting Squadron 12 (VF-12) had turned their machines over to Marines on Espirito Santo, Navy Fighting Squadron 17 (VF-17) "Skull and Crossbones" was the first Navy Corsair squadron to see action. Commanded by Tommy Blackburn, VF-17 became the first land-based fighter unit in the New Georgia area, and within 79 days of combat was credited with the destruction of 154 Japanese aircraft. This squadron has been called "the greatest Navy fighter squadron in history". It contained twelve aces (i.e.pilots credited with five or more victories) and destroyed no less than 18 torpedo bombers in two passes while providing top cover for the carriers Essex and Bunker Hill during the first strike on Rabaul. When they ran low on fuel VF-17 became the first squadron to "operate" from a ship in combat.

The Corsair achieved a victory/loss ratio of 11-3/1; it proved "definitely superior" in trials with a captured Zero and gave favorable results in competitive manceuvres with a P-51 , a P-47 , a P-38 and a P-39. Above 12,000 ft. the Corsair outfought the Mustang and was considered evenly matched at lower altitudes. Against the F6F (even with Lt.-Cdr. "Butch" O'Hare at the controls of the Hellcat) the Corsair was more than a match for its opponent.

The most important naval attack fighter of W.W.II, the Chance-VoughtCorsair remained in production for thirteen years, In all, about 12,600 Corsairs were built, up to 1952, in many versions ranging from F4U-1 to F4U-6 (AU-1). The Corsair has been credited with over 2,000 enemy airplanes destroyed. Not bad for an aircraft whose service trial, for its chosen role, ended in failure.

Ten Corsairs were ordered as the F2G-2 with the 3,000 hp Pratt and Whitney Wasp Major R-4360-4 28 cylinder engine. The R-4360 saw service only at the end of the war. This engine was used in late B-29s, which were actually B-50s, and the Convair B-36.

The F2G-2 was a naval versions of the F2G-1 with serial numbers 88495/88468 and used as carrier based, low altitude fighters. Maximum speed was 431 mph (694 km/h) at 17,000 ft (5,182 km). Armament consisted of four .50 calibre machine guns with an external bomb load of 3,200 lbs (1,452 kg).

After the war, Corsairs continued flying with several air forces, and became the final piston engined fighters built in the United States. When the Korean War started in 1950, Corsairs were again used by the US Marines for ground attack. Others supplied to the French Navy, in Indochina, remained in service until 1964. They continued to serve in Honduras, El Salvador and Argentina. Not until the mid-seventies did the last South American country finally withdraw the type from service.

Specifications:

Vought F4U-1D Corsair—Carrier-based, fighter-bomber

Dimensions:

Wing span: 40 ft 11 in (12.47 m)

Length: 33 ft 4 in (10.17 m)

Height: 16 ft 1 in (4.90 m)

Weights:

Empty: 8,694 lb (3,947 kg)

Operational: 12,039 lb (5,465 kg)

Maximum Take-Off: 13,120 lb (5,951 kg)

Performance:

Maximum Speed:

425 mph (684 km/h) @ 20,000 ft (6,096 m)

Cruise Speed: 182 mph (293 km/h)

Service Ceiling: 33,900 ft (10,333 m)

Normal Range:1,015 miles (1,633 km)

Maximum Range: 1,562 miles (2,514 km)

Powerplant:

One Pratt & Whitney Double Wasp ,

2,250 hp. R-2800-8W eighteen-cylinder radial engine.

Armament:

Six .50 calibre machine guns. External load of

2,000 lbs. (908 kg) bombs or eight rockets.

APPENDIX C

"Fred's Funnies"

by Fred "Sparks" Blechman

Although the following stories have nothing to do with flying Corsairs, or flying at all, they may be amusing to those of you that own computers.

"Fred's Funnies" also appears as an appendix in my recent book, "Simple, Low-Cost ElectronicsProjects," ©LLH Technology Publishing, with permission. To order this book, call 1-800-247-6553, or visit www.LLH-Publishing.com .

"CONFESSIONS OF AN ELECTRONIC GENIUS"

by Fred "Sparks" Blechman

This story is EXACTLY as written over 20 years ago for an amateur radio "ham" magazine called "73." Little has changed since, except transistors, integrated circuits, and digital electronics have added to the mystery. . . and now I'm called a "computer genius!" NOT!

Have you ever been asked to fix a single-sideband transmitter, even though you weren't really sure how a simple oscillator works? Well, I have. In fact, I'm always being asked questions I shouldn't be asked. Why? Because in the minds of some around me, despite my claims to the contrary, I am an electronic genius!

How did I achieve this status? How can you attain for yourself the dubious distinction of being an "electronic genius?" Well, if you promise not to blab it around, here's the story. . . .

The Genius Is Born

I suppose it all started when I decided to build my own radio-control equipment for a model airplane. The fact that I knew nothing about electronics didn't stop me; I was surrounded at work by electronic geniuses who could solve virtually any problem involving the lowly electron. Or so I thought.

Anyhow, the kit I bought was a real collection of mysterious goodies; wire, coils, tubes, phenolic, and those cute little cylindrical things with the pretty colored bands. I meticulously followed the instructions and sketches in the assembly of the receiver, a simple "single-tube super-regenerative receiver," according to the description.

Since I had no equipment to check out its operation, I took it to work for the electronic geniuses to fire-up. They performed their usual mystical rites with strange looking devices. The receiver refused to be impressed by the display. . . and just did not work!

The next two weeks were almost too painful to describe. Complete lunch hours were consumed in discussion, theory and testing by the geniuses. My greatest contribution was keeping my fingers crossed. The geniuses, individually and collectively, all had their chance at trying to seduce "Fred's Folly" into operation. Words like superheterodyne, intermediate frequency, converter, and mixer were generously sprinkled throughout their discussions. "But," I kept repeating, "this is a superregenerative receiver!"

The geniuses thought I had flipped. "Regenerative receivers went out with the Model T" they said, patting me on the head sympathetically.

Well, they finally gave up, and I was about to take up basket weaving as a new endeavor, when a hot spell proved fortunate. I noticed one of the silver-colored cartridge-shaped things in the receiver was leaking at one end, apparently from the heat. Could this be a bad part? It was marked ".01 MFD 100V." When this unit was replaced, the receiver worked. I had fixed it! The geniuses just shook their heads. "You are truly an electronic genius" they confided. . .

The Genius Grows

The bug had bitten. More receivers, more transmitters. . . and many more problems. Somehow, never really knowing how or why, I always managed to stumble on a solution. Pretty soon I found myself fixing other guys' equipment. You've heard the expression "the blind leading the blind". . .

About this time I decided to really find out what electronics was all about. Somehow I was not able to find anyone who was willing to sit down with me for twenty minutes and tell me all there is to know about electronics. So I attended night classes at the local high school, where I got to twirl knobs in the lab. I bought test equipment with knobs of my own to twirl. I repaired every radio the neighbors found in their attics. And, most important of all, I subscribed to 73 Magazine.

My reputation grew. Radio repairing is, after all, mostly tube changing, dial-cord restringing, replacement of obviously cooked parts, and a generous seasoning of good luck. (Knowing what you're doing can replace the good luck; in my case the good luck was the essential ingredient.) "You," they would tell me, "are an electronic genius!" By this time I was able to identify at least three different kinds of parts.

The Genius and the Theory

I found myself more and more becoming a victim of the never expressed, but universally accepted, theory of the masses: "He who knows anything about electronics knows everything about electronics." There is, however, a lesser known corollary to this theory: "He who knows anything about any particular branch of electronics knows practically nothing about any other branch of electronics!" I couldn't convince anyone that the latter theory more expressed my capabilities. "If it plugs into the wall, or uses a battery, Fred knows all about it," they insisted.

The Genius Takes to the Air

Then I got my ham ticket. That really did it! When my roof began sprouting weird antennas, and the neighbors' TV sets began acting in a strange manner, they were more convinced than ever that another Steinmetz was their private electronic consulting engineer. I was asked about everything from ailing TV sets (I carry service insurance on my own set) to improperly operating electric blankets (when mine quit recently, I bought a new one.) And it doesn't end there; I've even found myself answering questions on the air about how to plate-modulate a transmitter, or how to eliminate chirp on CW. Sometimes I have some idea what I'm talking about, but certainly not always. However, if I tell them I don't know what I'm talking about, then I am considered overly modest; if I offer no suggestions, the conclusion is that I don't care enough to even think about the problem. A dilemma. I have found it easier to give them an answer they don't understand than to try to convince them that I'm talking through my chapeau.

The Genius Goes Stereo

Take the other night, for instance. Andy, who has known me long enough to know better, brought over a stereo tape recorder he had just built from a kit. . . his first tussle with electronics. He said that the left channel was dead. Not being a tape recorder specialist, or any other kind of specialist, I did the only thing I could think of at the moment; I plugged in the "kluge" and turned it on. Music poured forth from both channels, loud and clear.

"What did you do to it?" Andy asked.

"Nothing," I replied.

"There you go being modest again," he said. "All you electronic geniuses are alike."

Then we tried to record. No erase. So I unbuttoned the whole works and looked at the maze of wire and stuff and things inside the chassis. I noticed two shielded cables from the erase head terminating in two plugs on the chassis. On a wild hunch (my usual method) I swapped the two plugs in their sockets. This cured the trouble. To Andy this was sheer wizardry. When I tried to explain the four-track stereo system, and the operation of the record and erase oscillator, he absorbed about as much as a third grader trying to learn the Pythagorian Theorem.

That's about the time the left-channel playback went dead. I had no recourse but to resort to the scientific approach. Using the dirty wooden handle of a small, dirty paintbrush that happened to be lying on my dirty workbench, I pushed and shoved everything in sight under the chassis. Responding to this precision trouble-shooting technique, the left channel burst forth in full bloom. More probing disclosed that a single strand of shielding had lodged itself against the grid of the left channel pre-amp tube!

Now the left-channel magic-eye record level indicator tube was acting oddly. Andy was obviously right-handed! No amount of pushing and shoving with the paintbrush handle did any good. This exhausted my supply of magic tricks, so I suggested that we put the whole works back in the case and be glad that it hadn't gone up in smoke. All buttoned up, we gave it the final check. No one was more surprised than I when everything worked, including the left-channel magic-eye indicator!

"You did something when I wasn't looking," accused Andy.

With a knowing expression, I replied, "The hand is quicker than the 'eye,' my friend. . . ."

"K.I.S.S. OR K.I.C.K.?"

by Fred "Kisser" Blechman

Many times in organizing a task, such as setting up a data file, designing a computer program, selecting new equipment, or managing your hard disk, you are faced with decisions that can either simplify or complicate the task. Do you K.I.S.S. it or K.I.C.K. it?

You have probably heard of the K.I.S.S. approach—"Keep It Simple, Stupid!" You may give this a lot of lip service, while you dream up bigger and better complexities. Instead of KISSing the job, you are probably KICKing it. K.I.C.K.? That stands for "Keep It Confusing, Knucklehead!"

After all, if you confuse everyone around you, it protects your job. They'll think you're so smart they can't get along without you. You become the local "guru" by being a KICKer rather than a KISSer.

I've been amazed over the years to see so many examples of KICKers in action—even outside the computer world. Most doctors, lawyers, accountants and engineers are KICKers rather than KISSers. Psychologists and psychiatrists are big KICKers. Probably the biggest KICKers of all are government workers and bureaucrats at all levels. How about politicians, who, after all, usually got their KICK training as lawyers?

In the musical world, in 1934 Cole Porter wrote "I Get a KICK Out of You", with "flying too high" as part of the lyrics. Apparently he changed his thinking by 1948 when he wrote the Broadway musical "KISS Me, Kate."

I was aware of the deep entrenchment of KICKers in the corporate structure when I worked in the aerospace industry as an engineer, surrounded by KICKers at all levels. Every time I tried to KISS, I'd get KICKed! Cost-plus-fixed-fee contracts don't benefit from KISSing, only KICKing.

However, when there is an incentive to KISS, complexity gets KICKed out. Businessmen—especially self-employed entrepreneurs—are usually KISSers. Why? Because they are dealing with their own bucks! The more they KISS, the bigger the "bottom line."

KISSing The KICKers

I taught myself electronics, many years ago, by reading Popular Electronics, Radio-Electronics, Electronics Illustrated and other electronic magazines (most of them gone today.) I was frequently appalled at the complexity of some projects that could be done with two or three transistors instead of the dozen used in some KICKer articles. So I started writing KISS articles. I became known for my KISSing.

Then I got into my own part-time Amway distributor networking business. I started KICKing until I saw how much time and effort I was wasting on needless tasks. As I saw negative cash flow, I became the biggest KISSer in town! The business became so successful, I retired from aerospace after five years—and haven't worked for anyone else in the 23 years since.

When the paperwork in our growing business got overwhelming, I bought a small computer and looked for programs for my wife to use. She's a great KISSer! But all the programmers I found were KICKers. They wanted to address the power of the computer instead of the simplicity of our needs. I KISSed them off and wrote my own programs.

My programs were simple to learn and simple to use—a KISSers delight. Originally written in 1980 for a 16K RAM TRS-80 Model I, eight programs address the needs of the business rather than the capabilities of a computer. The $49 "AMBIZ-PAK" has long since been improved and converted to the IBM PC.

How to KISS

I don't keep the prices of thousands of product prices in memory. That's too slow and awkward, so I KISS the top 400, and the program easily handles the others. I don't do inventory control; it's easier to look on the shelf. I don't keep track of each distributor's volume of business all month. It's faster to add the order totals on a calculator for the relative few that qualify for bonuses based on volume. I don't do double-entry bookkeeping or generate a profit and loss statement; I just keep putting money in the bank. The names and addresses of my distributors are still on 3 x 5 file cards. I don't know (and don't care) how many bottles or boxes of particular products I've sold in the last three years, by customer, by date, by amount, or by any other parameter. I'm a KISSer, and have better things to do than analyze things that don't matter. If I can't KISS it, I don't do it.

The "proof of the pudding" is that my wife, Ev, who HATES computers, uses these programs for distributor order processing and bonus statements, customer invoices, and bookkeeping—and now has TWO computers! She tells me that if it weren't for the computers and my simple programs, I would be doing the paperwork!

Wanna' Become A LOVER?

This whole KISS philosophy will probably shock some of you—especially those of you who use a $400 computer program with an 80486 PC/AT computer, 120 megabyte hard disk, and VGA color monitor to balance your check book. You may be using a powerful database program to keep track of your Christmas Card List or recipes, or a $500 word processing program and $2000 laser printer for your monthly letter to your parents. Would you buy a big truck to carry a bottle of milk from the store? Is it possible that's what you're doing with your computer?

Next time you're faced with a task, ask yourself a few questions. Is this the simplest way to do this job? Is this "overkill"? Isn't there an easier way? Would 3 x 5 cards be simpler? Would you rather be a KISSer or a KICKer? Remember, KISSers eventually become LOVERs ("Leave Out Various Extras, Rube!")

“MEMORIES ARE MADE OF THIS”

by Fred “Memory-Man” Blechman

I don’t know about you, but I’m confused about the various kinds of computer memory. Someone, somewhere seems to get some kind of thrill by creating names for memory (and most other computer-related hardware) that defy explanation. Furthermore, the names sound so much alike that more confusion results.

We’ve all heard of “RAM” (random access memory) and “ROM” (read-only memory.) Sound alike, don’t they? They even LOOK alike—integrated circuit chips that can only be told apart by their mysterious markings.

How about “expanded memory” and “extended memory?” After reading at least a dozen articles about these, I’ve decided that even the article authors don’t really know the similarities and differences between these—especially when software can apparently make one simulate the other!

So I’ve decided to add to the confusion with some new memory nomenclature. See if you can keep these in YOUR memory.

Expended Memory: This is the computer’s memory you are already using. Some is RAM, some is ROM, some is expanded, and some is extended.

Distended Memory: This is expended expanded extended memory beyond the address capabilities of the computer unless you add an 80386 coprocessor.

Unattended Memory: This is memory you’re not yet using, but available in your machine. Everything beyond expended expanded extended distended memory. May be used to run flight simulators in background mode while your boss is nearby.

Intended Memory: This is what you'll need to run Windows 3.0 and all future applications from Microsoft. You thought you could survive with 640K? Don't be ridiculous!

Pretended Memory: This is what you thought you had when you bought your computer—and later found out would require adding some expensive chips. Typically, this is having a four megabyte memory board in a machine with only 640K of RAM actually installed.

Tremendous Memory: Many gigabytes.

Stupendous Memory: Beyond tremendous.

Defended Memory: This has something to do with the 80386 "protected mode." Don't ask me; I don't even work here.

Contented Memory: This is the portion of memory where all your happy little TSR programs joyfully reside, snugly hidden away from your applications, but ready to jump into action to please you.

Frequented Memory: Your disk cashe stays here, often used and constantly on call.

Demented Memory: The part of memory reserved for thriller games like Dungeons and Dragons.

Suspended Memory: This is where your program is hiding when your computer "goes out to lunch."

Surrendered Memory: This is the memory you recover when you hit the RESET button.

Out-Of Memory: This is the no-memory land to which most of my programs seem to find their way.

And here are some more memory names for IBM to use, as yet undefined: enchanted, lamented, amended, offended, blended, fomented, and rescinded. Do you have some others?

"FRED'S TWENTY-FIVE MISTRESSES"

by Ev "Wife" Blechman

My husband, Fred, has had twenty-five mistresses. At least, that's all I know about. Of course, they say the wife is the last to know—but not in my case.

It happened! Not one, not two, not three, not four, but twenty-five times! And the sad thing is that I found out about the first one the moment he carried HER across our threshhold.

It was a cold January morning in 1978. Fred and I had only been married two months when he brought HER right to our doorstep. She was well disguised, but I was immediately intimidated by her because she weighed less than I did!

Fred carried her over the threshold with no explanation and then, just like the ravenous beauty that pops out of the cake at a bachelor's last gig, SHE popped out of her carefully designed outer garments.

Physically, my contours were much more curvy than hers. She was small, definitely angular, but solid. I should have deduced that from her outer garments. While I try to maintain a good California tan and a sunny disposition, she appeared in chic grey with black and silver accessories, and had a distinctive cool and precise manner.

The delicate way in which Fred handled—and fondled—her should have warned me of the constant confrontations which were to become part of our future, and which I've endured now for fifteen years. I admit to a slight scowl as I noticed Fred handling her so gingerly, and carefully, like a thing of great fragility. I didn't take his pulse, but it must have been racing as he picked her up, turned her over, inspected her closely, and gently ran his hands all over her frame.

She arrived, reminiscent of a bride and her trousseau, with various accessories. Fred immediately took charge, attaching the accessories in the right places. I couldn't help taking note of his excited manner, his joy, the wondrous expression of anticipation of glorious hours that he would spend with her. . . ..and I was jealous.

Because I was Fred's new bride, there was a strong obligation on Fred's time; I expected it to be spent with me! When we got married he said "Stick with me, Babe. We'll get rich building our Amway business together."

I won out for a short while. Then came the long, lonely hours waiting for him to leave HER side. I never knew when I would see the light go out in the guest room where he had made a home for her. Many times, late into the night, I would reach out to touch him, only to find that he was not there. However, he was always at my side when I woke up in the morning—but dead to the world!

I was beside myself. I had overheard Fred tell a friend on the telephone that he was teaching her to do just about anything he wanted. No wonder he appeared so haggard as he dragged himself into bed in the wee hours of the morning.

I quickly discovered a marked distinction between us. I didn't have all the answers, she did. And my Fred was trying to find them! I built up quite a resentment over this third party who was spending twenty-four hours a day in our new-bride paradise.

Something had to be done. I'd had it. Fred was being "wenched" away from me. One night I stomped my way towards the guest room and flung open the door. Just as I suspected, there was Fred, hovering over her as he had day after day for the last six months. She remained still—not a move. But I heard her humming to him. My presence was completely ignored. I hurled out of the room, slammed the door behind me and waited for some response. None. They were totally engrossed doing their thing together.

Hours later Fred came into our bedroom and announced he'd done all he could with her, and that he'd need to get another more powerful "model". I said nothing as my fury built up. He had made up his mind and my silence was mistaken for acquiescence. He got three more. All from the same family!

As time went by, Fred had twenty-five of these—you guessed it—microcomputers under our roof (although, thank goodness, not all at the same time!) Instead of being restricted to just the guest room, Fred used a total of three bedrooms, and spent more and more time running from one bedroom to the other. Over a period of time, some left as new ones arrived.

Just imagine how you would feel if you saw your home was invaded first by a TRS-80 Model I, then three Model IIIs, two Model 4s, two Model 4Ps, two Sinclair ZX-81s, a Timex Sinclair 1500, a Timex Sinclair 2068, a Radio Shack MC10, a Coleco ADAM, an Apple IIc, a Sanyo MBC 555s, a Sinclair QL, two Sinclair Spectrums, an IBM PC/XT Clone, a 286 clone, a booksize PC/XT, a Toshiba 1000 laptop, a Laser PC4 notebook, and a Microgold 286 portable. (Know any wife who would learn the names of her husband's twenty-five mistresses?)

At this point in time, there are only twelve micros left, and like a family, they share some of the bedrooms, the play room, and the office. It took a while to convince me that I would be able to accept and eventually love, if not all, one or two of Fred's "mistresses." Believe it or not, I can't wait to get my hands on one every day when I do paperwork for our Amway business—with the $49 "AMBIZ-PAK" of eight programs Fred wrote while I thought he was "fooling around." Maybe a wife IS the last one to know. . . . . .

"BLECHMAN'S TEN LAWS OF COMPUTING"

by Fred "Lawman" Blechman

I've owned 26 microcomputers, I've written over 500 magazine articles and five books specifically about microcomputer hardware and software since 1978—and I've come to the following conclusions, which I call "Blechman's Ten Laws of Computing":

(1) "When it's manufactured, it's already 'obsolete'—but still far more powerful than you need."
(2) "When you try to use it, it's incompatible with everything you have that used to work."
(3) "When you try to return it, they're out of business—or suddenly don't understand English."
(4) "No matter how big your hard drive, it will be filled within 30 days or less—mostly with things you'll never use."
(5) "Nothing works the first time, and never works when you try to show it off."
(6) "Everything you use is attacking your body with electromagnetic radiation of various sorts."
(7) "Any software upgrade costing less than $20 is an admission of guilt."
(8) "Version 1 of any software is full of 'bugs.' Version 2 fixes all the bugs and is great. Version 3 adds all the things users ask for, but hides all the great stuff in Version 2."
(9) "Any software costing over $100, or with documentation of over 100 pages, is too complicated."

(10) "You probably were better off with 3x5 cards and a typewriter in the first place!"

There are ways to "beat" these "laws"—but that could be the subject of a l-o-n-g article (and a 1-hour talk I've given to computer local clubs.) I only offer these "laws" here to counter the enormous computer hype that is avalanching unsuspecting buyers who might well be satisfied with plain-vanilla used PCs, XTs or 286 machines with monochrome monitors. In a recent check of local computer stores, I found none that sold less than 486 or Pentium machines with Super-VGA color monitors. This is gigantic overkill for many prospective users. Do you agree, or are you one who must have the latest and greatest—whether you need it or not, and regardless of how much it will complicate your life?

"WORD CERTAIN 2.0—THE ULTIMATE WORD PROCESSOR?"

by Fred "Last Word" Blechman

My friend, Mel Marcus, is a software freak! When he had his Commodore 64 years ago he collected hundreds of public domain and shareware programs. Then, when he got into the IBM PC world, he went nuts with shareware programs from computer bulletin board systems (BBSs) for years, downloading everything he could find.

Lately, like a growing number of cyber-freaks, Mel spends hours and hours roving around the electronic black hole known as The Internet, with its thousands of BBSs and "newsgroups" and their hundreds of thousands of downloadable files. So when Mel tells me he has found something new and exciting, I listen.

When Mel called me the other day and said "Fred, I've finally found it! I found it!" he got my attention.

"What did you find, Mel? I didn't know you were looking for anything."

Oh, yes!" he explained. "Ever since I started using microcomputers, I've been looking for a great word processor. I've found some good ones, but they were always too complicated. I do a lot of writing in my business, and I need something fast and simple.

"With the Commodore I used Paper Clip," he continued. "But since getting my first PC, I've found so many word processors to choose from, it's bewildering. Of course, there's WordStar and all its workalikes. I've tried MultiMate, VolksWriter, Leading Edge, WordVision, LeScript, The Word, WordPerfect, Word for Windows, and just about every one around. In the shareware world, I found PC Write to be about the most powerful.

"But," he continued, "they are all either too limited or too complicated. The manuals are hundreds of pages long. Printer installations can drive you bananas. And just about the time you learn how to use one of these word processors, they put out a new version that won't read the old text files! And the Windows versions—well, they confuse me altogether. So many buttons, bars, icons, and dialogue boxes, I get totally lost and can't find the things I used to use in DOS. . . "

"I know what you mean," I replied. "But you have finally found one you like? Tell me about it."

"Well", Mel started out, "it's called Word Certain 2.0. I've never seen it advertised, but I ran across it on the Internet. Best of all, it's not even shareware—it's public domain, so you don't even have to pay anyone to use it! It was written by a guy named Kevin Mitnick. I've got it on disk. You can come over and run the disk yourself. It's fabulous! I even wrote out a list of features that. . . .."

"Hold it," I interrupted. "Can you send me that list right away? The deadline for the next issue of Nuts & Volts is coming up. Maybe I can tell the world about Word Certain 2.0."

"Sure, Fred. I'll drop it in the mail tonight."

A couple of days later I received the list from Mel:

WORD CERTAIN 2.0 Main Features
(Compiled by Mel Marcus)

1. Instant access. No waiting for DOS to load.
2. All commands internal. No accessing disk for commands or overlay programs. No extended memory required.
3. 270K storage capacity (about 45,000 words, or 180 double-spaced pages.)

4. Each character immediately stored to reduce chance of data loss.
5. Instant error correction with included utility.
6. Full search and replace functions.
7. Thesaurus and dictionary compatible.
8. Desktop publishing ability lets user specify layout, and freely create and import graphics. Automatic hand-scanning and image editing.
9. Prints out in selected colors.
10. Simple, easily understood commands.
11. Produces upper and lower case English and foreign characters, as well as proof-reading symbols.
12. Prints in many different fonts and sizes. Especially efficient for script fonts.
13. Does not require special printer installation or paper feed mechanism. Works with pin-feed or cut sheet paper.
14. Subscripts, superscripts, underlining, bold, enhanced.
15. Insert, delete, indents and other formatting functions.
16. Cut and paste for block moves.
17. Page numbering, headers, and footers.
18. Automatic on-line real-time spelling and grammar checker.
19. Import from Lotus, Dbase and other popular programs.
20. Can be modified by user without learning new language.
21. Easily transportable between different computers.
22. Does not conflict with memory-resident programs.
23. Multitasking. Can be used while computer is running a background program.
24. Data easily duplicated and shared with other users.
25. Can also create spreadsheets, forms, and data bases with no additional software.
26. Works with Mac or PC without modification.

I was impressed. I called Mel.

"Mel, you say you have Word Certain on disk. Can you send me a copy? I'd like to see for myself. I find it hard to believe. Can it really do all those things on your list?"

"Hey, Fred, it's unbelievable! Everyone has been after me for copies and it's been driving me crazy. I don't make a nickel on this. I'm out of blank disks and mailers. Why don't you come over and I'll let you run the disk?"

My curiousity was at a peak. This sounds like a natural article for Nuts & Volts. Larry and Robin love new stuff! Mel's place is only a few miles from me, so I made an appointment with him that afternoon.

When I got to Mel's door I recognized the yellow pad and pencil he always had on his door for notes. There was a note for me: "Fred: Had to leave for a few minutes. Come on in. The computer is on and ready."

I went inside and there was Mel's 486 multimedia monstrosity. On the screen it said, "Type DIR, press Enter, then look at README.1ST." I typed DIR and pressed the Enter key. The screen showed a bunch of files like WC.EXE, WCINSTAL.EXE, WCHELP.DOC and README.1ST. I typed TYPE README.1ST and pressed Enter. The screen went blank then displayed this message:

LOOK AT THE CALENDAR ON THE WALL, THEN TURN AROUND.

I looked at the calendar. It said April 1. I turned around to see Mel standing there, grinning from ear to ear, and holding out a yellow pad and a slim, yellow pencil with a nice rubber eraser on top.

"Here's Word Certain 2.0. Like I told you, it's unbelievable. See," he said, holding out the pencil, "the top is software, the wooden part is hardware, and the point is the printer. The world's simplest word processor. . ."

He's probably right! How can you beat a pencil and paper for simplicity?

INDEX